Virus Culture

The Practical Approach Series

SERIES EDITOR

B. D. HAMES
School of Biochemistry and Molecular Biology
University of Leeds, Leeds LS2 9JT, UK

See also the Practical Approach web site at **http://www.oup.co.uk/PAS**

★ indicates new and forthcoming titles

Affinity Chromatography
Affinity Separations
Anaerobic Microbiology
Animal Cell Culture (2nd edition)
Animal Virus Pathogenesis
Antibodies I and II
Antibody Engineering
★ Antisense Technology
Applied Microbial Physiology
Basic Cell Culture
Behavioural Neuroscience
Bioenergetics
Biological Data Analysis
Biomechanics – Materials
Biomechanics – Structures and Systems
Biosensors
Carbohydrate Analysis (2nd edition)
Cell-Cell Interactions
The Cell Cycle
Cell Growth and Apoptosis
★ Cell Separation
Cellular Calcium
Cellular Interactions in Development
Cellular Neurobiology
★ Chromatin
★ Chromosome Structural Analysis
Clinical Immunology
Complement
★ Crystallization of Nucleic Acids and Proteins (2nd edition)
Cytokines (2nd edition)
The Cytoskeleton
Diagnostic Molecular Pathology I and II
DNA and Protein Sequence Analysis
DNA Cloning 1: Core Techniques (2nd edition)
DNA Cloning 2: Expression Systems (2nd edition)
DNA Cloning 3: Complex Genomes (2nd edition)

DNA Cloning 4: Mammalian Systems (2nd edition)

★ DNA Microarrays

★ Drosophila (2nd edition)

Electron Microscopy in Biology

Electron Microscopy in Molecular Biology

Electrophysiology

Enzyme Assays

Epithelial Cell Culture

Essential Developmental Biology

Essential Molecular Biology I and I

★ Eukaryotic DNA Replication

Experimental Neuroanatomy

Extracellular Matrix

Flow Cytometry (2nd edition)

Free Radicals

Gas Chromatography

Gel Electrophoresis of Nucleic Acids (2nd edition)

★ Gel Electrophoresis of Proteins (3rd edition)

Gene Probes 1 and 2

Gene Targeting

Gene Transcription

★ Genome Mapping

Glycobiology

★ Growth Factors and Receptors

Haemopoiesis

★ High Resolution Chromotography

Histocompatibility Testing

HIV Volumes 1 and 2

★ HPLC of Macromolecules (2nd edition)

Human Cytogenetics I and II (2nd edition)

Human Genetic Disease Analysis

★ Immobilized Biomolecules in Analysis

Immunochemistry 1

Immunochemistry 2

Immunocytochemistry

★ *In Situ* Hybridization (2nd edition)

Iodinated Density Gradient Media

Ion Channels

★ Light Microscopy (2nd edition)

Lipid Modification of Proteins

Lipoprotein Analysis

Liposomes

Mammalian Cell Biotechnology

Medical Parasitology

Medical Virology

★ MHC Volumes 1 and 2

★ Molecular Genetic Analysis of Populations (2nd edition)

Molecular Genetics of Yeast

Molecular Imaging in Neuroscience

Molecular Neurobiology

Molecular Plant Pathology I and II

Molecular Virology

Monitoring Neuronal Activity

Mutagenicity Testing

★ Mutation Detection

Neural Cell Culture
Neural Transplantation
Neurochemistry (2nd edition)
Neuronal Cell Lines
NMR of Biological Macromolecules
Non-isotopic Methods in Molecular Biology
Nucleic Acid Hybridisation
Oligonucleotides and Analogues
Oligonucleotide Synthesis
PCR 1
PCR 2
★ PCR 3: PCR *In Situ* Hybridization
Peptide Antigens
Photosynthesis: Energy Transduction
Plant Cell Biology
Plant Cell Culture (2nd edition)
Plant Molecular Biology
Plasmids (2nd edition)
Platelets
Postimplantation Mammalian Embryos
★ Post-translational Modification
Preparative Centrifugation
Protein Blotting
★ Protein Expression
Protein Engineering
Protein Function (2nd edition)
Protein Phosphorylation
Protein Purification Applications
Protein Purification Methods
Protein Sequencing
Protein Structure (2nd edition)
Protein Structure Prediction
Protein Targeting
Proteolytic Enzymes
Pulsed Field Gel Electrophoresis
RNA Processing I and II
★ RNA-Protein Interactions
Signalling by Inositides
★ Steroid/Nuclear Receptor Superfamily
Subcellular Fractionation
Signal Transduction
★ Transcription Factors (2nd edition)
Tumour Immunobiology
★ Virus Culture

Virus Culture

A Practical Approach

Edited by

ALAN J. CANN

Department of Microbiology and Immunology,
University of Leicester

UNIVERSITY PRESS

OXFORD
UNIVERSITY PRESS

Great Clarendon Street, Oxford OX2 6DP

Oxford University Press is a department of the University of Oxford and furthers the University's aim of excellence in research, scholarship, and education by publishing worldwide in

Oxford New York

Athens Auckland Bangkok Bogotá Buenos Aires Calcutta Cape Town Chennai Dar es Salaam Delhi Florence Hong Kong Istanbul Karachi Kuala Lumpur Madrid Melbourne Mexico City Mumbai Nairobi Paris São Paulo Singapore Taipei Tokyo Toronto Warsaw

and associated companies in Berlin Ibadan

Oxford is a registered trade mark of Oxford University Press

Published in the United States by Oxford University Press Inc., New York

A catalogue record for this book is available from the British Library

Library of Congress Cataloging in Publication Data
(Data available)

ISBN 0-19-963715-6 (Hbk)
0-19-963714-8 (Pbk)

Typeset by Footnote Graphics,
Warminster, Wilts
Printed in Great Britain by Information Press, Ltd,
Eynsham, Oxon.

Preface

The properties of viruses are distinct from those of living organisms, which makes the study of virology different from other areas of biology. Many specialized techniques have been developed to study viruses, and yet these have almost invariably found their way into mainstream biology. It is not possible to define in a single volume the techniques of virology, but this volume and its companions set out to illustrate the major experimental methods currently employed by virologists.

When I was first approached by David Hames to edit this volume, my initial reaction was to decline. However, on brief reflection I decided that the Practical Approach series was important enough to the research community that I should take my turn as an editor. Having agreed to edit *Virology: a practical approach*, it was only during the course of our subsequent discussions that the magnitude of what I had acquiesced in fully dawned on me. My thanks go to David Hames (joint series editor) for his guidance in shaping this and the accompanying volumes. Inevitably there will be those who feel that this or that should have been included or left out. It is not possible to include everything within the format of this series!

Most importantly, thanks must go to the contributors who were prepared to share their combined expertise with the wider research community. Chapter 1 provides an overview of cell culture, essential for all virologists, and frequently forgotten by those with a tendency to slip into bad habits. Chapters 2 and 4 are complementary insights into the fundamentals of virology: isolating and identifying viruses. Chapter 3 describes methods for concentrating and purifying viruses, Chapter 5 is on electron microscopy as applied to virology, and Chapter 7 describes antiserum and monoclonal antibody production. Chapter 6 is unusual for a Practical Approach volume in that it is rather more theoretical that most. The reason for this is that it is difficult to cover all approaches to virus vaccines in detail, but my feeling is that this is such a significant area of virology that a reasonably comprehensive treatment deserved to be included. This chapter is complemented by the methods for antiviral drug testing described in Chapter 8. Finally, the last chapter describes the specialized techniques used to study plant viruses, which will not be of direct relevance to all readers, but is important to an overall understanding of practical virology, and the impact that molecular biology has had on this field.

Finally, a word of advice to students (and others!) who experience difficulties with any of the protocols. If you cannot get a protocol you know to have worked previously to work for you, there are two choices. Either troubleshoot the whole protocol in a step-by-step fashion with reference to appropriate controls, in order to identify and rectify problems, or use a different protocol. In my experience, many experimental problems arise from mutilated protocols.

These procedures are sequential and cumulative, where each step is dependent on the previous ones. It is *not* a good idea to insert steps from one protocol into another, or to omit steps, until you have already performed the procedure successfully several times. That said, all of the protocols in this volume are tried and tested.

Leicester A. J. C.
May 1999

Contents

Contributors

S. BIEL
Robert Koch-Institut, Nordufer 20, D-13353, Berlin, Germany. (biels@rki.de)

K. BLAKE
European Collection of Animal Cell Cultures, Centre for Applied Microbiological Research, Porton Down, Wiltshire SP4 0JG, UK.

A. J. CANN
Microbiology & Immunology, University of Leicester, University Road, Leicester LE1 7RH, UK. (nna@le.ac.uk)

S. DAVIES
Retroscreen Ltd, 64 Turner Street, London E1 2AD, UK.

H. R. GELDERBLOM
Robert Koch-Institut, Nordufer 20, D-13353 Berlin, Germany. (gelderblom-@rki.de)

J. J. GRAY
Public Health Laboratory, Level 6, Addenbrookes Hospital, Hills Road, Cambridge CB2 2QW, UK. (jg2@mole.bio.cam.ac.uk)

W. IRVING
Department of Microbiology, Queens Medical Centre, Nottingham NG7 2UH, UK. (will.irving@nottingham.ac.uk)

R. JENNINGS
Sheffield Institute for Vaccine Studies, Division of Molecular and Genetic Medicine, Floor F, University of Sheffield Medical School, Beech Hill Road, Sheffield S10 2RX, UK. (r.jennings@sheffield.ac.uk)

L. S. KELLY
Retroscreen Ltd, 64 Turner Street, London E1 2AD, UK.

R. LAMBKIN
Retroscreen Ltd, 64 Turner Street, London E1 2AD, UK. (rlambkin@retroscreen.com)

S. LENNOX
Department of Microbiology, University of Cape Town, Private Bag, Rondebosch 7700, South Africa.

J. MCKEATING
Department of Microbiology, University of Reading, London Road, Reading RG1 5QA, UK. (j.a.mckeating@reading.ac.uk)

Contributors

P. D. MINOR
NIBSC, South Mimms, Potters Bar EN6 3QG, UK.

C. W. POTTER
Sheffield Institute for Vaccine Studies, Division of Molecular and Genetic Medicine, Floor F, University of Sheffield Medical School, Beech Hill Road, Sheffield S10 2RX, UK (r.jennings@sheffield.ac.uk)

J. S. OXFORD
Retroscreen Ltd, 64 Turner Street, London E1 2AD, UK.

E. P. RYBICKI
Department of Microbiology, University of Cape Town, Private Bag, Rondebosch 7700, South Africa. (ed@molbiol.uct.ac.za)

C. SHOTTON
Hybridoma Unit, McElwain Laboratories, Institute of Cancer Research, Sutton, Surrey SM2 5NG, UK. (chriss@icr.ac.uk)

A. STACEY
European Collection of Animal Cell Cultures, Centre for Applied Microbiological Research, Porton Down, Wiltshire SP4 0JG, UK.

M. VALERI
Hybridoma Unit, McElwain Laboratories, Institute of Cancer Research, Sutton, Surrey SM2 5NG, UK.

Abbreviations

ACV	acyclovir
ATCC	American Type Culture Collection
ATP	adenosine triphosphate
BMS	black Mexican sweetcorn suspension cells
BSA	bovine serum albumin
cfu	colony-forming units
CMV	human cytomegalovirus (HHV-5)
CPE	cytopathic effect
CuMV	cucumber mosaic virus
DEAFF	detection of early antigen fluorescent foci
DEPC	diethylpyrocarbonate
dH_2O	(sterile) distilled water
DMSO	dimethyl sulfoxide
ECACC	European Collection of Cell Cultures
EDTA	ethylenediaminetetraacetate
EID_{50}	egg infective dose 50
ELISA	enzyme-linked immuno sorbant assay
EM	electron microscopy
EthBr	ethidium bromide
FA	formaldehyde
FBS	fetal bovine serum
FTM	faecal transport medium
GA	glutaraldehyde
GMP	good manufacturing practice
HBSS	Hanks balanced saline solution
HIV-1	human immunodeficiency virus type 1
HIV-2	human immunodeficiency virus type 2
IEM	immuno-electron microscopy
IL2	interleukin-2
mAb(s)	monoclonal antibody(ies)
MDCK	Madin Darby dog kidney
MEM	minimum essential medium
mesh	number of grid holes per inch
MOI	multiplicity of infection
NPA	nasopharyngeal aspirate
PBG	phosphate-buffered gelatin
PBMC	peripheral blood mononuclear cells
PBS	phosphate-buffered saline
PBSA	phosphate-buffered saline/ 0.1% (w/v) sodium azide
PCR	polymerase chain reaction

pfu	plaque forming units
PTA	phosphotungstic acid
RBC	red blood cells
RLV	Rauscher leukaemia virus
RNP	ribonucleoprotein
RSV	respiratory syncytial virus
SEM	scanning electron microscopy
SPIEM	solid phase immuno-electron microscopy
STA	silicotungstic acid
$TCID_{50}$	tissue culture infectious dose 50
TEM	transmission electron microscopy
TK	thymidine kinase
TMV	tobacco mosaic virus
UAc	uranyl acetate
UF	uranyl formate
UTM	urine transport medium
VTM	viral transport medium

1

Cell culture

K. BLAKE and A. STACEY

1. Introduction

The ability to culture cells *in vitro* has enabled enormous advances to be made in biology, particularly in virology. Many cell types can be routinely proliferated and cryopreserved, allowing consistent, characterized, contaminant-free cells to be used both as a research and diagnostic tool. However, despite the utility of cell culture, researchers must recognize that cells grown *in vitro* may not behave as cells in the intact organism. For example, the proliferation of cells *in vitro* is not necessarily conducive to full differentiation, where cell growth is normally absent. The potential loss of differentiated characteristics will inevitably affect properties and therefore the utility of cells. Cell behaviour will also be affected by the loss of cell–cell, and cell–matrix interactions, and loss of tissue architecture. The goal of attaining a greater range of *in vivo* characteristics is becoming more achievable with the increased availability of growth and differentiation factors, and the development of three-dimensional culture systems which allow cells to assume their correct polarities, shapes and interrelationships. The future perspective looks promising, and cell culture systems may be becoming available which enable *in vitro* growth and study of currently non-cultivable viruses (Chapter 2).

This chapter outlines techniques for the growth, quantification, and cryopreservation of cell cultures. In addition, important quality control and authentication issues are addressed.

1.1 Sources of cell cultures

Cells derived from a broad range of species and tissue types may be maintained or propagated *in vitro* (1, 2, 3). With the development of many different media and growth factors, the range of cultivable cell types is increasing. Examples are given below.

1.1.1 Vertebrate mammalian

Techniques are available (4) to cultivate normal or tumour cells from the following vertebrate mammalian (predominantly human or rodent) tissue systems:

- Integument and muscular, e.g. melanocytes, keratinocytes, muscle and fat cells
- Gastro-intestinal, e.g. salivary epithelium, pancreas, liver, colon cells, smooth muscle cells
- Respiratory, e.g. lung, alveolar and bronchial epithelium
- Urinary, e.g. kidney cells
- Female reproductive, e.g. amniocyte, chorionic villi, cervical cells, etc.
- Male reproductive, e.g. Sertoli cells
- Endocrine, e.g. thyroid, pancreas cells
- Osteo-articular, e.g. chondrocytes, synovial cells, osteoblasts
- Neuronal, e.g. dorsal root ganglion, neuronal, glial cells
- Cardiovascular, e.g. cardiac myocytes, umbilical cord vein endothelial cells
- Haemopoietic, e.g. progenitor cells, macrophages, T and B cells

Other mammals from which cell cultures have been derived include rabbit, guinea pig, horse, cattle, dog, cat, gerbil, monkey, racoon, etc.

1.1.2 Vertebrate non-mammalian

Cell cultures may be derived from vertebrate non-mammalian species, including fish, amphibia, and birds (4). Fish cell lines have been derived from a range of species, predominantly from embryo tissue, originally for the growth and study of fish viruses. Culture techniques and media are those commonly used for mammalian cell culture, although fish cells have a reduced requirement for glutamine and may be incubated at temperatures lower than 37°C, e.g. at room temperature.

1.1.3 Invertebrate

Invertebrate cell culture is required for baculovirus growth. Again, techniques for the maintenance of cultures are similar to those used for mammalian cell culture. Insect cells grow at a lower temperature (27°C) than mammalian cells, and require specialized media.

1.2 Suppliers of cell cultures

It is strongly recommended that cell cultures are obtained from recognized collections such as the European Collection of Cell Cultures (ECACC) or American Type Culture Collection (ATCC) (*Table 1*). These organizations provide pure, authentic, quality-controlled cell cultures with information concerning appropriate media and cultivation conditions. An example of data provided with each cell line obtained from ECACC is given below (*Table 2*).

Table 1. Suppliers of cell cultures

Culture collection	Address
ECACC	European Collection of Cell Cultures Centre for Applied Microbiology & Research, Salisbury, Wiltshire SP4 0JG, UK http://www.camr.org.uk/ecacc.htm/
ATCC	American Type Culture Collection P.O. Box 1549, Manassas, Virginia 20108, USA http://www.atcc.org/
DSMZ	Deutsche Sammlung von Mikroorganismen und Zellkulturen GmbH Mascheroder Weg 1B, D-3300 Braunschweig, Germany http://www.dsmz.de/
Riken	Riken Cell Bank 3-1-1 Koyadai, Tsukuba Science City, 305 Iboraki, Japan http://www.rtc.riken.go.jp/

Table 2. European Collection of Animal Cell Cultures – cell line data sheet

CELL NAME:	**Vero**	**ECACC No:**	**84113001**
DESCRIPTION:	Monkey African Green Kidney		
Culture Medium:	DMEM + 2 mM glutamine + 10% fetal bovine serum		
Sub–culture routine: cells per cm^2 using 0.25%	Split confluent cultures 1:3 to 1:6 i.e. seeding at 1–3 × 10 000 trypsin or trypsin–EDTA; 5% CO_2, 37°C.		
Cell Type: Adherent	Passage No: 144	Lot No:	Cells/ml: 2×10^6
Expected Viability:	90%	Resuscitate into: 1 × $25cm^2$ flask	
References:	Nippon Rinsho 1963; 21:1209		

1.3 Types of cell culture

In theory, tissues can be taken from any organism and primary cell cultures obtained. If desired, these primary cultures (1.3.1) can be subcultured (i.e. moved into a separate culture vessel, preferably following dilution to expand the culture). Some cultures, more often fetal in origin, can be serially passaged, but will eventually reach senescence; these are termed finite cell lines (1.3.2). Other cultures may spontaneously undergo crisis, be derived from tumour material, or be exposed to transforming agents and produce continuous cell lines which have the capacity (in theory) to be subcultured indefinitely (1.3.3).

1.3.1 Primary cell culture

Tissue is taken from an organism, disaggregated if necessary, and placed into a culture vessel under suitable conditions of media, temperature, etc. The culture is termed 'primary' up to the first subculture. Primary cell cultures are the most representative of normal tissues, and are still valuable in some areas of virology, particularly for virus isolation. However, they are labour-intensive to obtain, and are heterogeneous in nature. Batch-to-batch variability, potential for contamination, and poor characterization are inevitable.

1.3.2 Finite cell lines

Following passage of primary cell cultures, some cultures will survive serial passage before becoming senescent and increasingly slow growing. Cell lines in this category obtained from normal healthy tissue will continue to display a 'normal' phenotype (*Table 3*). Examples of finite lines include widely used cells such as WI38 and MRC-5. Finite cell lines originally derived from embryonic tissues will generally have the potential for a greater number of population doublings before the onset of senescence than those derived from adult tissues.

Table 3. Characteristics of a "normal" phenotype and transformed phenotype, generally expressed by finite and continuous cell lines respectively

"Normal" phenotype	Transformed phenotype
Anchorage dependent	Reduced anchorage dependence and possible growth in suspension
Contact inhibition of growth	Loss of contact inhibition
May display differentiation	Usually undifferentiated
Diploid	Heteroploid, aneuploid
Finite lifespan	Infinite lifespan Reduced requirement for serum/growth factors. Shorter population doubling time. May form tumours on injection into animals.

1.3.3 Continuous cell lines

These include transformed and immortalized cell lines, and are derived directly from tumour material or by exposure of cells to transforming agents. Transformation of cells may be due to:

- Exposure to chemical carcinogens
- Ionising radiation
- Infection with retroviruses or DNA tumour viruses (or viral components)
- 'Spontaneous'

Following transformation, continuous cell lines express an altered phenotype (*Table 3*). Immortalization is one step in the transformation process, and may confer an infinite life span but without the alteration in growth control. Examples of continuous cell lines include Vero and HeLa cells.

2. Culture media

The purpose of cell culture media is to provide nutrients in a readily accessible form at optimal pH and osmolarity for cell survival and proliferation. In addition, the temperature, and the oxygen and carbon dioxide content of the cultures must be controlled. In general, all culture media consist of a basal medium plus serum and/or other growth factors.

2.1 Basal media

There are a variety of basal media for mammalian cell culture. The most widely used are Eagle's medium and derivatives, and RPMI medium and derivatives, with others available for particular cell lines and serum-free culture. A number of basal media are also available for insect cell culture, e.g. Mitsuhashi and Maramorosch's, Schneider's, and Grace's media. The major components and their roles are described in *Table 4*. For non-production scale operations, it is recommended that 1x liquid media is purchased ready for use, to avoid problems of sterility, water quality, etc. It may be necessary to add glutamine and other unstable components to liquid media prior to use. Media containing bicarbonate as a buffering system will require the addition of 5%

Table 4. Components of basal media and their function

Component	**Function**
Balanced salt solution	Maintain physiological pH, maintain osmotic pressure, membrane potential, cofactors for enzymes
Buffering systems e.g. Bicarbonate/CO_2, Hepes	Compensate for CO_2 and lactic acid production; HCO_3 is also a growth factor
Carbohydrates or glutamine e.g. glucose, galactose	Energy source
Amino acids	Essential amino acids not synthesised by cells, non–essential amino acids which may be lost by cells into medium
Vitamins e.g. para-aminobenzoic acid, biotin, folic acid, B_{12}, riboflavin etc.	Precursors for cofactors
Hormones and growth factors e.g. insulin, hydrocortisone, nerve growth factor, epidermal growth factor, fibroblast growth factor, etc.	Stimulate cell proliferation or differentiation.
Proteins and peptides e.g. fetuin, α–globulin, fibronectin, albumin, transferrin	Carry hormones, vitamins, lipids, etc.
Fatty acids and lipids	Membrane biosynthesis, etc.
Accessory factors e.g. trace elements, nucleotides	Enzyme co–factors, etc.

CO_2 in air as the gaseous phase in culture vessels. Recognized suppliers of cell cultures (ECACC, ATCC) will provide information concerning choice of culture medium and additives.

2.2 Serum

Fetal bovine serum (FBS) is the most commonly used supplement to basal media. It is a rich source of growth factors, which are effective across a broad range of cell types derived from various species and tissue types. The major components of serum and their function are detailed in *Table 5*.

Table 5. Main components of serum and their function

Component	Function
Growth Factors	Stimulate cell proliferation or differentiation.
Albumin	Carrier protein for small molecules e.g. lipids, steroids, vitamins, metal ions pH buffer; protects cells against mechanical damage in agitated systems
Transferrin	Iron transport
Anti–proteases, e.g. α_1 antitrypsin, α_2 macroglobulin	Prevent proteolytic damage to cells.
Attachment factors, e.g. fibronectin, fetuin, laminin	Allow binding of attachment dependent cells to substrate.

Serum may vary in performance from batch to batch, and should be tested prior to purchase. The country of origin of serum may be important if cells are to be imported at some later stage into the USA. Restrictions will be imposed on the importation of cell lines grown in serum obtained from countries with indigenous infections, including bluetongue, foot-and-mouth disease, and infectious bovine rhinotracheitis.

FBS is a major source of viral contaminants in cell culture, particularly bovine viral diarrhoea virus. Such viruses may be present in low numbers, and require sensitive detection methods such as PCR. Bovine serum is not thought to be a source of the BSE infectious agent, although it is advisable to avoid bovine sources for preparation of materials for human *in vivo* use.

2.3 Serum-free media

The development of serum-free media has been driven by the shortcomings of serum as outlined above, and the desire to simplify downstream purification of cell products. In general, there has been success in the development of serum-free media for particular cell types. Such commercially available media consist of basal media supplemented with components such as insulin, transferrin, growth factors, hormones, etc. However, there are no universally applicable serum-free media.

2.4 Antibiotics

The use of antibiotics should be saved for emergency decontamination of irreplaceable cultures, or for the preparation of primary material from non-sterile sites. Routine antibiotic use in culture media is not advised due to the potential for suppression of low-level bacterial contamination, and selection of antibiotic-resistant strains.

3. Good practice in cell culture

Good practice in cell culture can be reduced to three facets:

- Preparation of cell banks from good quality source material to provide low passage back-up stocks
- Maintenance of sterility
- Prevention of cross-contamination with other cell cultures

As described above, it is strongly recommended that cell cultures are sourced from recognized culture collections. On receipt, master cell-banks should be prepared, to provide stocks for future use in case of excessive passage levels or contamination. The risk of microbial contamination can be minimized by good working practices, staff training and adequate facilities. Good working practices include:

- Routine screening of cell cultures for the presence of contaminants such as bacteria, fungi and mycoplasma at least once a month.
- Handling of contaminated material in an area remote from contaminant-free culture, or if this arrangement is not possible, handling cell cultures in the following sequence, with inter-procedure disinfection of work surfaces, etc.:
  (a) known uncontaminated material
  (b) untested or unknown
  (c) known contaminated material
- Not using antibiotics routinely, as this can generate antibiotic-resistant strains and mask low levels of contamination.
- Handling of one cell culture at a time, and using a fresh bottle of medium for each cell line to reduce the risk of cross-contamination.
- Rooms and equipment should be cleared regularly.

Staff should be trained in aseptic technique as a minimum requirement, with further training in all aspects of cell culture available from culture collections, etc. It is recommended that cell cultures are handled in a Class II Microbiological Safety Cabinet which offers a degree of protection to both product and operator. A higher category of containment may be required if cell cultures are known to harbour ACDP Hazard Group 3 pathogens (5).

4. Principles of cell banking

Continuous propagation of cell cultures can lead to the loss of differentiated characteristics present in low passage cell cultures. Implementation of a master and working cell stock system (*Figure 1*) can help to overcome this, by storing material at a low passage level. Once a cell culture has been established, or obtained from another source such as a culture collection, the cultures should be expanded to allow several vials of cryopreserved material to be prepared in one batch, the size of which will be dependant upon the envisaged level of usage. This batch should then be examined using the quality control (QC) techniques described below to confirm the origin of the material, and to verify the absence of adventitious agents. This batch then serves as the master stock, from which a series of working stocks may be prepared. Each working stock should also undergo basic QC testing to verify that it is contaminant-free, and to establish conformity with the master stock.

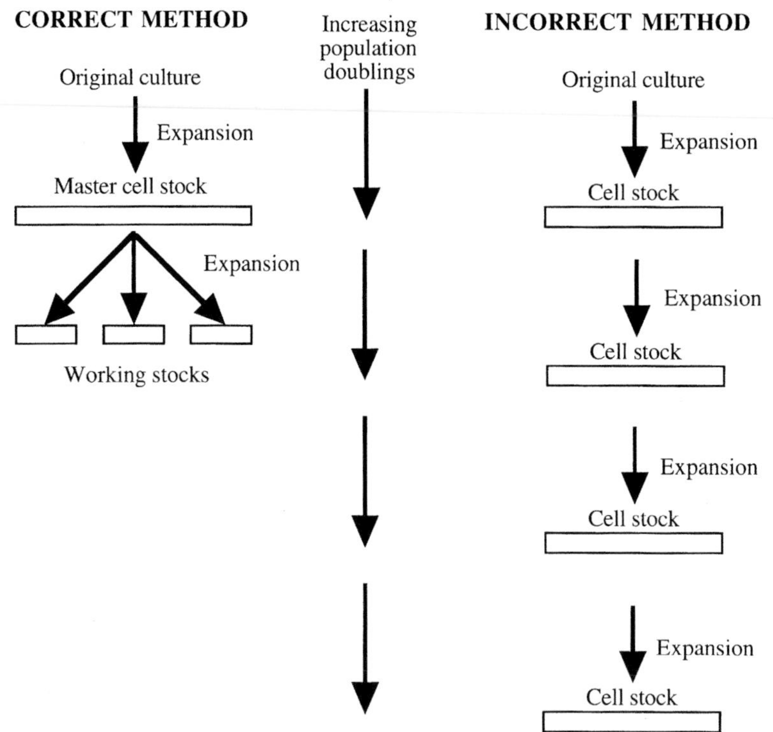

Figure 1. Comparison of correct and incorrect approaches to cell banking systems.

The major advantage of employing a system such as this is that all experiments are conducted on cells of similar passage and population doubling, which may be critical if the cells have a finite lifespan in culture. In addition, this gives a desirable level of standardization and reproducibility. The direct comparison of the master and working cell stocks by either karyotyping or DNA fingerprinting demonstrates the stability of the material during culture, and eliminates the risk of mis-identification of cultures. Ideally this comparison should be performed on material which is specifically kept in culture for a long period (up to 6 months' continuous culture) to confirm the long-term stability.

5. Subculture of cells

As living material, cell cultures require nutrients which are replenished by subculture. Cell cultures derived from different tissues and organs will have different morphologies and different growth characteristics, and therefore may require different methods of subculture. Cultures can take one of three basic forms:

- anchorage-dependent, growing as an adherent/attached monolayer
- in suspension
- semi-adherent monolayer (containing cells loosely attached to the plastic support, with a proportion of cells in suspension)

Protocols 1, 2, and 3 describe how to subculture attached, suspension, and semi-adherent cell lines respectively.

Protocol 1. Subculture of adherent cell lines.

Equipment and reagents

- Inverted microscope, preferably with phase-contrast optics (e.g. Leica, Nikon)
- Phosphate-buffered saline (PBS) (Sigma, Life Technologies, ICN Flow, etc.)
- 0.25% trypsin, or 0.05% trypsin-0.02% EDTA solution (Sigma, Life Technologies, ICN Flow, etc.)
- Fresh medium supplemented with fetal bovine serum (FBS), and other growth factors as required (Sigma, Life Technologies, ICN Flow, etc.)
- Fresh culture vessels (e.g. Corning, Falcon, Costar, etc.)

Method

1. Check the culture for the degree of confluence, morphology, and the absence of microbial contamination, using an inverted microscope (objectives ×10, ×20). Phase-contrast optics make details of cell morphology much easier to determine.
2. All solutions should be prewarmed to at least room temperature, not used directly from the refrigerator.

Protocol 1. *Continued*

3. Decant the medium and wash the cell layer twice with PBS, using approximately 10 ml per 25 cm^2 of surface area.
4. Add trypsin-EDTA (approximately 1–2 ml per 25 cm^2 surface area). Spread the trypsin over the cell layer and decant the excess leaving a 'film' of trypsin over the cell monolayer.
5. Incubate at 37 °C until the cells are detached from the flask (approximately 5–10 min). Check that they are all loose using an inverted microscope.
6. Wash the cells into the corner of the flask with fresh medium (2–3 ml per $25cm^2$).
7. Pipette gently to break-up any clumps.
8. If a viable cell count is necessary, e.g. when preparing cells for cryopreservation, take a sample (100–200 μl) and perform a cell count (*Protocol 4*)
9. Inoculate fresh flasks or roller bottles using the recommended seeding density and split ratios, adjusting the cell concentration with fresh medium accordingly.
10. Incubate under the appropriate conditions.

Protocol 2. Subculture of suspension cell lines

Equipment and reagents

- Inverted microscope, preferably with phase contrast optics (e.g. Leica, Nikon)
- Fresh culture vessels (e.g. Corning, Falcon, Costar, etc.)
- Haemocytometer (e.g. Sigma)
- Fresh medium supplemented with FBS and other growth factors as required (Sigma, Life Technologies, ICN Flow, etc.)

Method

1. Check the culture for morphology and absence of microbial contamination using an inverted microscope (objective ×10, ×20).
2. Gently tap the flask and check that all the cells are in suspension. Pipette gently to break up any clumps.
3. Perform a viable cell count if required (*Protocol 4*).
4. Inoculate fresh flasks or roller bottles as above.
5. Incubate under appropriate conditions.

Protocol 3. Subculture of semi-adherent cell lines

Equipment and reagents

- Inverted microscope, preferably with phase contrast optics (e.g. Leica, Nikon)
- Phosphate buffered saline (PBS) (Sigma, Life Technologies, ICN Flow, etc.)
- 0.25% trypsin, or 0.05% trypsin-0.02% EDTA solution (Sigma, Life Technologies, ICN Flow, etc.)
- Fresh medium supplemented with FBS and other growth factors as required (Sigma, Life Technologies, ICN Flow, etc.)
- Fresh culture vessels (e.g. Corning, Falcon, Costar, etc.)

Method

1. Check the culture for morphology and absence of microbial contamination, using an inverted microscope (objective ×10, ×20).
2. Decant the medium and non-adherent cells into a labelled centrifuge tube.
3. Wash the cell layer twice with PBS, using approximately 10 ml per $25 cm^2$ of surface area.
4. Add the trypsin-EDTA (approximately 1–2ml per 25 cm^2 of surface area). Spread the trypsin over the cell layer and decant the excess leaving a 'film' of trypsin over the cells.
5. Incubate at 37°C until the cells are detached from the flask (approximately 5–10 min). Check that they are all loose using an inverted microscope.
6. Wash the cells into the corner of the flask with fresh medium (2–3 ml per 25 cm^2).
7. Pipette gently to break up any clumps. Add the cells to the decanted medium and non-adherent cells in the centrifuge tube.
8. Perform a viable cell count as required (*Protocol 4*).
9. Inoculate fresh flasks or roller bottles as above.
10. Incubate under appropriate conditions.

6. Quantification of cell cultures

For many applications, the starting cell concentration is an important variable which needs to be standardized to give reproducibility of results. Numerous methods have been developed to achieve this, ranging from manual systems involving the use of a haemocytometer chamber (*Protocol 4*), to highly automated procedures such as flow cell technology. Automated systems undoubtedly give quick and reliable results; however, they require considerable financial investment. Conversely, manual cell counts are still quick and cheap to perform, but may be a little more subjective. Ultimately the choice of technology will be based upon the proposed level of usage.

Protocol 4. Quantification of cells using a standard or improved Neubaur haemocytometer

Equipment and reagents

- Inverted or conventional microscope, with long focal length ×10 or ×20 objective lenses (e.g. Leica, Nikon)
- Phosphate-buffered saline (PBS) (Sigma, Life Technologies, ICN Flow, etc.)
- 70% (v/v) ethanol in dH_2O
- 0.25% trypsin, or 0.05% trypsin-0.02% EDTA solution (Sigma, Life Technologies, ICN Flow, etc.)
- 0.49% (w/v) trypan blue in dH_2O (e.g. Sigma)
- Haemocytometer (e.g. Sigma)

Method

1. Bring the cell culture into suspension by trypsin treatment (*Protocols 1 and 3*), or by gentle pipetting (*Protocol 2*), as necessary.
2. Thoroughly clean the haemocytometer chamber and coverslip with 70% ethanol before use, and allow to dry.
3. Breathe onto the chamber to provide moisture and gently move the coverslip back and forth over the chamber until Newton's rings are visible.
4. Remove a small aliquot of cell suspension (100–200 μl) and mix with an equal volume of 0.4% trypan blue.
5. Take the cell-trypan blue mixture up into a pipette tip and gently rest at the junction between the chamber and the coverslip. The mixture will be taken up by the capillary effect.
6. Allow a few seconds for the cells to settle.
7. Record the number of viable cells (translucent) and non-viable cells (stained blue by trypan blue). At least 200 cells should be counted to ensure an accurate cell count.
8. Count one area of 1 mm^2 (quadrant) from both halves of the chamber.
9. Calculate total number of cells, the number of viable cells per ml, and the percentage viability, using the following equations:

Total number of cells/ml =

$$\frac{\text{Total number of cells counted} \times \text{dilution factor}}{\text{Number of 1mm}^2\text{ areas counted}} \times 10^4$$

$$\text{Viable cells/ml} = \frac{\text{Number of viable cells counted} \times \text{dilution factor}}{\text{Number of 1mm}^2\text{ areas counted}} \times 10^4$$

$$\text{\% viability} = \frac{\text{Number of viable cells/ml}}{\text{number of cells/ml}}$$

7. Cryopreservation and storage of cell lines

In order for cell cultures to retain their unique characteristics and to reduce the risk of genetic drift and microbial contamination, it is advisable to limit the period for which any given cell line is in culture. This may be achieved by the implementation of a master cell banking system (*Figure 1*) in conjunction with a tried and tested protocol for cryopreservation. Moreover, the costs of maintaining a cell line continuously are also reduced.

To achieve optimal long-term storage of mammalian, insect, or plant tissue, a few basic principles should be observed. Only viable, healthy cells should be frozen, at a cell concentration in the range $2–6 \times 10^6$ cells per ml. At cell concentrations greater then 10^7 cells per ml, the viability on recovery decreases. Ideally cultures should be in the log phase of growth to ensure maximum recovery rates. It is important to use a cryoprotectant suitable for the cell line in question. One of the most commonly used is dimethyl sulfoxide (DMSO); however, DMSO can have toxic effects on cells if not kept ice cold. Moreover, DMSO is not suitable for all cell types, since it induces differentiation in some human lymphoid cell lines, notably HL 60. Alternative cryoprotectants include sucrose and glycerol. All cryoprotectants have the same mode of action, in that they cause a rise in osmolarity in the medium, causing water to leave the cells. At the same time, the protectant coats the cells preventing the cell membrane from being pierced by any water crystals that form inside the cell as the temperature falls.

The practice relating to the cryopreservation of cell lines is to freeze slowly and thaw quickly. The easiest way to achieve a slow rate of cooling is to use a controlled-rate programmable freezer, which allows the user to design a programme to maximize cell recovery. In general terms a rate of cooling of 1 °C per minute is suitable for many cell types. Much research has been undertaken to achieve routine methods that will give success for cell lines derived from a range of species and tissues, and a basic protocol is given below (*Protocol 5*). A simple, effective and easy-to-use alternative is an isopropanol bath, such as the Mr Frosty system (BDH Chemical Co). In this system, vials of cells are placed within the isopropanol bath and incubated at −80 °C overnight. The contents should then be stored permanently in gas-phase liquid nitrogen.

Protocol 5. Cryopreservation of cell cultures

Equipment and Reagents

- Low speed centrifuge (e.g. Sorvall, Beckman)
- Controlled rate freezer (e.g. Planar) or isopropanol bath (e.g. Mr Frosty system, BDH Chemical Co)
- Haemocytometer (e.g. Sigma) or alternative method of counting cells
- Phosphate-buffered saline (PBS) (Sigma, Life Technologies, ICN Flow, etc.)
- 0.22 μm syringe filter units (e.g. Sartorius)
- 0.25% trypsin, or 0.05% trypsin-0.02% EDTA solution (Sigma, Life Technologies, ICN Flow, etc.)
- Fetal bovine serum (FBS) (e.g. Sigma, Life Technologies, ICN Flow)
- DMSO or glycerol (e.g. Sigma)
- Freezer vials (e.g. Nunc)

Protocol 5. *Continued*

Method

1. Record the details of the cell lines to be frozen, including the passage number, cell count, viability, media used, etc. Label the vials.

A. *Attached cell lines*

A1. Decant the medium, saving up to 30 ml in a centrifuge tube.

A2. Wash the cell sheet twice with PBS using approximately 10 ml per 25 cm^2 of surface area.

A3. Add trypsin-EDTA (approximately 1–2 ml per 25 cm^2 of surface area). Spread the trypsin over the cell layer, and decant the excess leaving a 'film' of trypsin over the cells.

A4. Pour off the excess trypsin, and incubate at 37 °C until the cells detach from the flask (approximately 5–10 min).

A5. Resuspend the cells in culture media from A1.

B. *Suspension cell lines*

B1. Aliquot the cells into centrifuge tubes.

C. *Semi-adherent cell lines*

C1. Decant the media and non-adherent cells into a labelled centrifuge tube.

C2. Wash the cell layer twice with PBS, using approximately 10 ml per 25 cm^2 of surface area.

C3. Add the trypsin-EDTA (approximately 1–2 ml per 25 cm^2 of surface area). Spread the trypsin over the cell layer and decant the excess leaving a 'film' of trypsin over the cells.

C4. Incubate at 37 °C until the cells are detached from the flask (approximately 5–10 min). Check that they are all loose using an inverted microscope.

C 5. Wash the cells into the corner of the flask with fresh medium (2–3 ml per 25 cm^2).

C6. Pipette gently to break-up any clumps. Add the cells to the decanted medium and non-adherent cells in the centrifuge tube.

2. Perform a cell count.
3. Centrifuge the cell suspension at 150 *g* for 5 min.
4. Prepare a solution of 91% FBS and 9% cryoprotectant (0.22 μm filtered DMSO or glycerol). Allow 1 ml for each vial. Store on ice.
5. Pour off the supernatant from the cell pellet following centrifugation.

Resuspend the cell pellet in the required volume of cryoprotectant solution, adjusting the cell concentration within the range 2–6 $\times$ 10^6 cells per ml.

6. Aliquot 1 ml of cell suspension into each vial, and tighten tops.
7. Place the vials in the freezing equipment.

7.1 Storage of cryopreserved cell stocks

For long-term storage (i.e. for periods of longer then 1–2 months) the material should be stored in vapour-phase liquid nitrogen. Storage at temperatures such as −20 °C and −80 °C can result in the loss of cell viability with time. Vials should not be stored in the liquid phase of nitrogen, due to the risk of ingress of liquid nitrogen and vials exploding. In addition, there is some degree of controversy surrounding the possibility of cross contamination of bacteria and mycoplasma between vials stored in liquid phase.

7.2 Resuscitation of cryopreserved cell stocks

As mentioned above, in order to achieve the maximum level of recovery from cryopreserved stocks, material should be frozen slowly and then thawed quickly, ideally in a waterbath set at 37 °C. In addition, the cryoprotectant used should be diluted quickly, or removed by centrifugation to minimize any potential toxic effects.

Protocol 6. Resuscitation of cryopreserved cells

Equipment and Reagents

- Protective clothing, i.e. lab coat, protective face mask, and gloves (e.g. Jencons)
- Ethanol (e.g. BDH)
- Low-speed centrifuge (e.g. Sorvall, Beckman)
- Fresh culture vessels (e.g. Corning, Falcon, Costar, etc.)
- Inverted microscope (e.g. Leica, Nikon)

Method

1. Wear protective clothing. Remove the required vial(s) from the liquid nitrogen tank, place the vial in a screw-top container within a transport tin, and remove to the laboratory.
2. Still wearing protective clothing, place the vial into a 37 °C waterbath until thawed (usually 3–5 min). Ensure that the vial cap is above the water level.
3. Immediately transfer the thawed vials to a microbiological safety cabinet, and open using an alcohol drenched tissue.
4. Slowly add the cells dropwise into a centrifuge tube containing 9 ml of medium, pre-warmed to 37 °C.

Protocol 6. *Continued*

5. Centrifuge the cells at 150 *g* for approximately 5 min at room temperature.
6. Pour off the supernatant and resuspend the pellet of cells in the appropriate medium. Seed cells into tissue culture flasks.
7. Incubate under the appropriate conditions.
8. After 24–48 h incubation examine the cells microscopically for characteristic growth and absence of microbial contamination. Subculture as required.

8. Quality control testing of cell lines

The quality of cell cultures in terms of purity, identity, and freedom from adventitious agents is important in scientific research; it is critical when they are used as substrates for diagnosis and production of therapeutic or vaccine material. The extent of quality-control testing undertaken depends on the ultimate use of the cell culture. If derived products are for *in vivo* use, then appropriate regulatory authorities (e.g. US Food and Drug Administration (FDA) (6, 7); US Animal and Plant Health Inspection Service, Department of Agriculture (8)) will determine the extent and methodology of quality-control tests to be employed. Testing of cell cultures for the presence of common adventitious agents should be routine in any tissue culture facility. Bacterial and fungal contamination can be detected by microscopic and sometimes by macroscopic examination. However, the detection of mycoplasma and virus contaminants requires the use of specific test procedures (7).

Contamination can exert numerous effects, for example, bacterial and yeast contamination will cause nutrient depletion, resulting in the death of the cell culture. Mycoplasma exert insidious effects on cell cultures, such as the alteration of the growth rate of cells, the induction of chromosomal aberrations, changes in amino acid and nucleic acid metabolism, and membrane aberrations. Techniques such as isoenzyme analysis and karyotyping can be used routinely to establish the species of origin (9). Moreover, techniques such as DNA fingerprinting (10) not only allow the identification of a cell line (11) but also enable the genetic stability of the material during routine and extended culture periods to be studied (12, 13). Isoenzyme and cytogenetic analysis are authentication methods recognized by regulatory authorities such as the FDA (7) and may be required when applying for a product licence.

8.1 Preparation of cell cultures prior to testing

Prior to testing for any adventitious agent, cell cultures should undergo at least two passages in antibiotic-free medium, since contamination may be suppressed by the presence of antibiotics. Equally, cryopreserved stocks should also undergo two passages in antibiotic-free medium due to the inhibitory effects of cryoprotectants. Suspension cell lines may be used direct. However, cell lines that grow as an attached monolayer should be brought into suspension, using a standard method of subculture, with trypsin and/or EDTA. Cells should be resuspended in the original cell culture medium at a cell concentration of approximately 5×10^5 cells per ml.

8.2 Mycoplasma

8.2.1 Sources of contamination

Historically, animal serum and trypsin were recognized as likely sources of mycoplasma; however, the screening of tissue culture reagents is now more rigorous. The most common source of mycoplasma is other cell lines, where contamination is spread by aerosols created during the work processes. The frequency of mycoplasma infection in some tissue culture laboratories may be as high as 100%. The most common contaminants are *Mycoplasma hyorhinis, Acholeplasma laidlawii, Mycoplasma orale, Mycoplasma fermentans,* and *Mycoplasma arginini.*

8.2.2 Methods for detection of mycoplasma

As mentioned above, mycoplasma are not visible with the naked eye or microscopically. Therefore specific tests are required to confirm their presence or absence. Direct DNA staining is one such technique (*Protocol 7*). Cultures are inoculated on to sterile coverslips, placed in tissue culture dishes and incubated for 24–72 h. Cultures are then fixed and stained with a pan-DNA stain such as DAPI or Hoechst 33258, and examined at ×100 magnification under UV epifluorescence. Cell nuclei will fluoresce. In mycoplasma-negative cultures the nuclei will be seen against a dark background. In mycoplasma-positive cultures, the cell nuclei will be seen amongst fluorescing thread-like or coccal structures (*Figure 2*).

Protocol 7. Detection of mycoplasma by DNA staining

Equipment and reagents

- UV microscope (e.g. Zeiss), filter set (excitation 265λ, beam splitter 395λ, emission 420λ)
- Carnoy's fixative (3:1 (v/v) methanol-glacial acetic acid). Prepare 4 ml of fixative for each sample to be tested. N.B.: Care must be taken when disposing of used fixative.
- Hoechst stain stock solution: Dissolve 10 mg bisbenzimide Hoechst 33258 (Sigma) in 100 ml of distilled water. Filter sterilize using a 0.2 µm filter unit (Sartorius). Wrap the container in aluminium foil and store in the dark at 4°C. The toxic properties of Hoechst 33258 are unknown, therefore gloves should be worn at all times when handling the powder or solutions.
- Hoechst stain working solution: Add 50 µl of stock solution to 50 ml of distilled water. Prepare immediately before use.
- Mountant: 22.2 ml 1 M citric acid, 27.8 ml 0.2 M disodium phosphate; autoclave at 15 lb in^{-2} for 15 min, and then mix with 50 ml glycerol; adjust to pH 5.5; filter sterilize, and store at 4°C.

Positive controls:

- *Mycoplasma orale* strain NCTC 10112
- *Mycoplasma hyorhinis* NCTC 10130

Protocol 7. *Continued*

Method

1. Add 2–3 ml of cell suspension to each of two tissue culture dishes containing autoclave-sterilized glass coverslips.
2. Inoculate positive control dishes in duplicate with 100 colony forming units (cfu) of *M. orale* and *M. hyorhinis*. Additionally one pair of dishes should be left uninoculated as a negative control.
3. Incubate at 36°C in a humidified 5% CO_2-95% air atmosphere for 12–24 h.
4. Remove one dish, and incubate the remaining dish for a further 48 h.
5. Before fixing, examine the cells for the absence of bacterial or fungal contamination. Do not decant the medium.
6. Fix the cells by adding 2 ml of Carnoy's fixative dropwise at the edge of the dish to avoid disturbance. Leave at room temperature for 3 min.
7. Carefully remove the fixative and tissue culture medium to a waste bottle, and add a further 2 ml of fixative to the dish. Leave for 3 min.
8. Pipette the fixative to waste.
9. Invert the lid of the dish, and using forceps rest the coverslip against the lid for 10 min to air dry.
10. Wearing gloves, return the coverslip to the dish and add 2 ml Hoechst stain (working solution).
11. Shield the coverslip from direct light and leave at room temperature for 5 min.
12. Pipette the stain to waste.
13. Add one drop of mountant to a labelled slide and place the coverslip cell-side down.
14. Examine the slide at ×100 magnification with oil immersion under UV epifluorescence.

In an alternative indirect system, cells of the test culture can be inoculated on to coverslips pre-inoculated with an indicator cell line such as the Vero African Green Monkey cell line. In this case the Vero cells are be inoculated on to the coverslips 4–24 h prior to addition of the test cell lines. The major advantage of this system, which compensates for the additional time required, is the increased sensitivity achieved by the increased surface area of cytoplasm in Vero cells, which aids visualization of mycoplasma. This system also enables the mycoplasma screening of serum and other reagents that can be inoculated directly onto the indicator cell line.

Cell culture samples can be inoculated directly onto mycoplasma agar

(a)

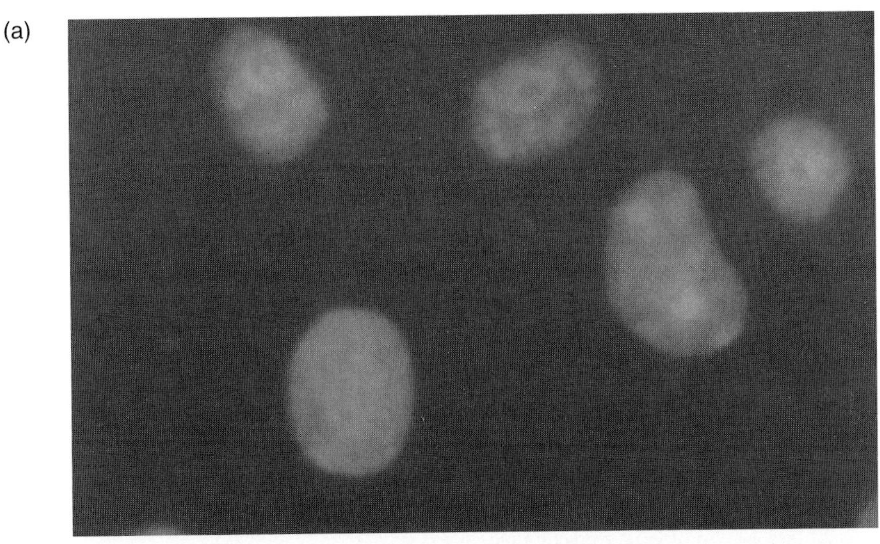

(b)

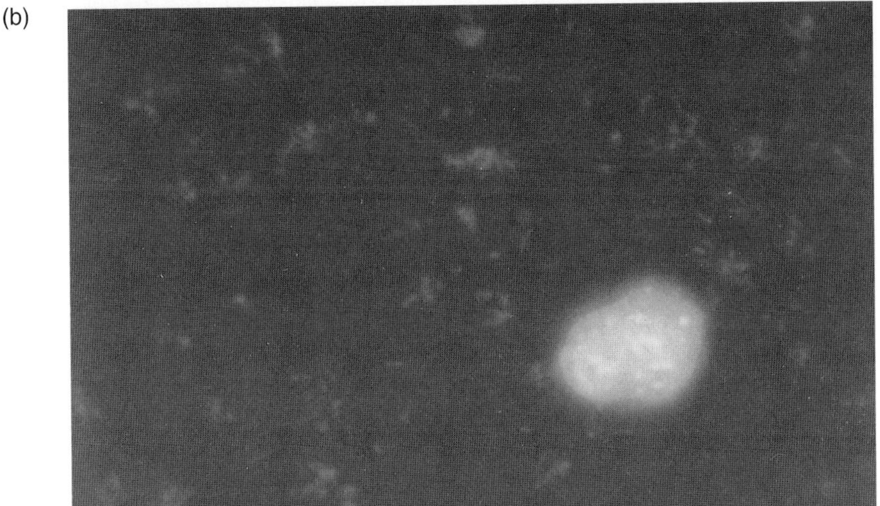

Figure 2. (a) DNA-stained mycoplasma-free cell culture. (b) DNA-stained mycoplasma-contaminated cell culture.

plates, supplemented with pig serum and yeast extract, and incubated at 36 °C ± 1 °C for 14–21 days, after which time they are examined, using an inverted microscope, for the presence of mycoplasma colonies. In addition, the original sample also undergoes an enrichment step in mycoplasma broth, supplemented with horse serum and yeast extract, for 7–14 days prior to inoculation onto mycoplasma agar plates. This simple step can increase the sensitivity of the assay. Mycoplasma colonies are visible on the agar within 2–7 days, depending upon species. Typically, mycoplasma colonies will be spherical, and some species produce colonies that have a distinctive 'fried egg' appearance (*Figure 3*). In some instances it may be difficult of distinguish true mycoplasma colonies from 'pseudocolonies' and cell aggregates. The use of Dienes stain, which stains true mycoplasma colonies blue, but leaves pseudocolonies and fungal and bacterial colonies unstained, can be used. Additionally, by using a sterile bacteriological loop, cell aggregates can be disrupted, but mycoplasma colonies will leave a central core embedded in the agar.

Protocol 8. Detection of mycoplasma by cultivation

Equipment and reagents

- Agar preparation:
  (a) Dissolve 2.8 g of mycoplasma agar media base (Oxoid) in 80 ml distilled water, and autoclave at 15 1b in^{-2} for 15 min. Store at 4°C for up to 3 weeks.
  (b) Dissolve 7 g of yeast extract (Oxoid) in 100 ml distilled water, and autoclave as above. Using aseptic technique dispense into 10 ml aliquots and store at 4°C for up to 1 month.
  (c) Using aseptic technique, dispense pig serum (Imperial Laboratories) into 10 ml aliquots and heat inactivate by incubation of serum at 56°C for 45 min. Store at 4°C for up to 6 months.
  (d) Allow the autoclaved agar media base to cool to 50°C and mix with 10 ml of heat-inactivated pig serum and 10 ml yeast extract (both pre-warmed to 50°C).
  (e) Dispense 8 ml per 5 cm diameter Petri dish. Seal in plastic bags and store at 4°C for up to 10 days before use.
- Broth preparation:
  (a) Dissolve 2 g of mycoplasma broth base (Oxoid) in 70 ml distilled water and autoclave at 15 1b in^{-2} for 15 min. Store at 4°C for up to 3 weeks.
  (b) Dissolve 7 g of yeast extract (Oxoid) in 100 ml distilled water and autoclave as above. Dispense into 10 ml aliquots and store at 4°C for up to 1 month.
  (c) Dispense horse serum (Tissue Culture Services) into 20 ml aliquots. Store at −20°C for up to 6 months. N.B. Do not heat inactivate.
  (d) Allow the autoclaved agar media to cool to 50°C and mix with 20 ml of horse serum and 10 ml yeast extract (both pre-warmed to 50°C).
  (e) Dispense 1.8 ml per glass vial, and store at 4°C for up to 1 month.
- Gas jars (Gallenkamp)
- Gas-pak anaerobic system and catalyst (Gallenkamp)
- Positive Controls: *Mycoplasma orale,* NCTC 10112; *Mycoplasma pneumoniae,* NCTC 10119

Method

1. Inoculate an agar plate with 0.1 ml of the test cell suspension, and incubate anaerobically at 36°C for 21 days.
2. Inoculate two agar plates as positive controls with 100 colony forming units (cfu) *M. orale* and *M. pneumoniae*, and incubate as above. Additionally, leave one plate uninoculated as a negative control and incubate as above.
3. Inoculate a broth with 0.2 ml of the test cell suspension and incubate aerobically at 36°C.
4. At approximately 7 and 14 days post-inoculation, subculture 0.1 ml of the inoculated broth cultures onto fresh agar plates and incubate as above in 1.

5. After 7, 14 and 21 days incubation, the agar plates should be examined, under ×40 or ×100 magnification with an inverted microscope, for the presence or absence of mycoplasma colonies.

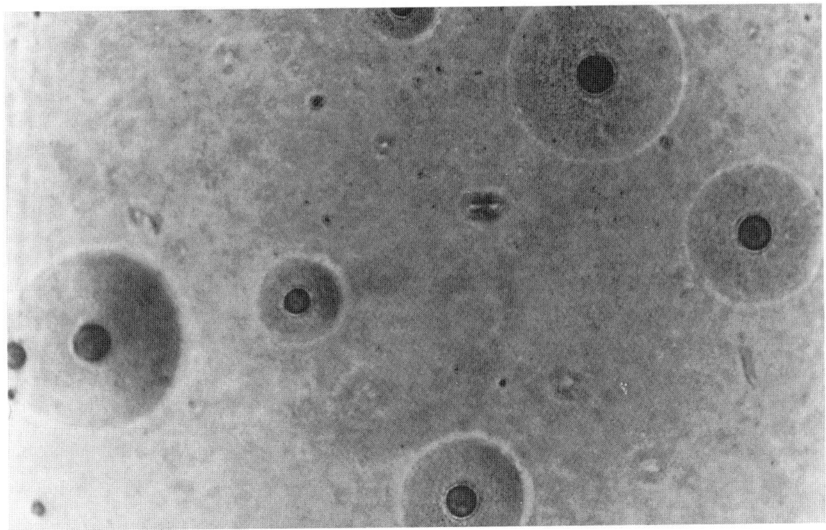

Figure 3. Mycoplasma colony growing on agar displaying 'fried-egg' morphology.

8.2.3 Detection of mycoplasma using commercially available kits

i. Gen-Probe mycoplasma TC rapid detection system

This system (Eurogenetics) employs the principle of nucleic acid hybridization to detect mycoplasma and acholeplasma RNA in tissue culture. A mycoplasma/acholeplasma RNA-specific DNA probe bound to beads is incubated with tissue culture supernatants at 72°C, causing cell lysis and allowing the probe to bind to mycoplasma RNA if present. Unbound probe is removed by washing. Bound probe is detected by means of the [^{3}H] label present on the probe, by the use of a scintillation counter. Using this system it is possible to detect positive samples in three hours or less.

ii. Boehringer Mannheim mycoplasma detection ELISA assay

This detection system is based on the ELISA technique. ELISA plates are coated with polyclonal anti-mycoplasma species antiserum, and then incubated with either tissue-culture supernatant, or positive and negative controls. Following the removal of unbound antigen, bound antigen-antibody complexes are detected using species-specific biotin-labelled antibodies visualized by binding of avidin-alkaline phosphatase. Using this system it is possible to not only confirm the presence of mycoplasma but also identify the species.

However, it should be noted that only four species are detected, namely *M. orale, M. hyorhinis, M. arginini* and *A. laidlawii.*

iii. Mycoplasma PCR ELISA

The mycoplasma PCR ELISA assay (Boehringer Mannheim) is a two-step process, involving PCR using mycoplasma-specific primers, followed by an ELISA designed to detect PCR products. This gives a reported sensitivity of 10^3 cfu ml^{-2}, equivalent to 1–10 fg DNA (1 to 20 gene copies) per PCR reaction. However, this detection level is dependent upon species of mycoplasma. PCR reactions are set up using a cell pellet obtained by centrifugation of 1 ml of culture supernatant. Following lysis and neutralization of the cell pellet, the PCR is set up using reagents provided, standard cycling conditions and digoxigenin-labelled UTP (DIG). The PCR product is then rendered single-stranded, hybridized with a biotin-labelled capture probe, and immobilized on streptavidin-coated wells of an ELISA plate. Bound probe/product is then visualized following incubation with anti-DIG antibody conjugated to horseradish peroxidase and addition of enzyme substrate.

iv. MycoTect®

The MycoTect® detection method (Life Technologies) is an enzymatic method, using the enzyme adenosine phosphorylase, which is found in differing amounts in mycoplasma and mammalian cells. In the assay, the enzyme catalyses 6-methylpurine deoxyriboside (6-MPDR), an adenine analogue, to methylpurine and 6-methylpurine riboside, both of which are toxic to mammalian cells. Thus, a mycoplasma-infected culture can be detected by incubation with 6-MPDR and monitoring for cell death, for a period of 4–5 days.

Most of the above tests will detect all species of mycoplasma, and closely related organisms such as *A. laidlawii. Table 6* shows a comparison of the methods described. Isolation by culture will detect all species, except for *M. hyorhinis* which is non-cultivable. The commercial kits detect all species with varying degrees of sensitivity, with the exception of the ELISA method which detects four species only (*M. orale, M. hyorhinis, M. arginini,* and *A. laidlawii*). Poor sensitivity of a detection method may result in low-level contamination being missed. The tests described generally show a threshold of detection of at least 10^4 cfu ml^{-2} of sample. Greater sensitivity can be obtained with the culture method, which has a theoretical detection threshold of 1 cfu ml^{-2} of sample for cultivable species. However, in order to achieve this level of sensitivity a culture period of 4 weeks is necessary. Also, great sensitivity is not possible with commercially available media. In addition, cultures should be tested every time they are recovered from nitrogen storage, since mycoplasma may proliferate more quickly than animal cells.

The DNA staining and culture methods have a relatively low cost per test, as compared with the commercial kits; however, the interpretation of results requires a significant level of training and experience. The MycoTect® system also has a low cost per test and requires little training; however, results are not

Table 6. Comparison of different techniques used for the detection of mycoplasma infection

Test	Sensitivity	Detection range	Speed	Regulatory authority approval	Special equipment requirements
DNA stain	10^3–10^4 cfu	All species	1 day	FDA[a] only	UV microscope
Culture	1 cfu	All cultivable species (excludes *M. hyorhinis*)	2–4 weeks	EP[b] and FDA[a]	None
Gen-probe	10^3–10^4 cfu	All species	1 day	none	Scintillation counter
ELISA	10^3–10^4 cfu	4 species only	2 days	none	ELISA reader
MYCOTECT	10^3 cfu	All species	4–5 days	none	none

[a] FDA – Food and Drug Administration (USA)
[b] EP – European Pharmacopoeia

obtainable for 4–5 days. A further consideration is the need for specialist equipment. As can be seen from *Table 6*, most of the tests described (except for the culture method and the MycoTect® system) require specialist equipment, which adds to the set-up cost. On final analysis, the choice of technique used will be based largely upon the availability of skilled staff and the frequency of testing required. For a larger number of tests, DNA-staining techniques will probably represent the most cost effective way of providing a comprehensive screening system, provided adequate training of staff. Training for all of the tests described is available through culture collections, which can also provide mycoplasma testing services to regulatory approved standards.

8.2.4 Mycoplasma eradication

In the event of cultures becoming infected with mycoplasma, the best course of action is to discard the cultures and, following extensive decontamination of the tissue culture cabinets and work surfaces, resuscitate 'clean' cell stocks. However, in the case of irreplaceable stocks this may not be practicable. In cases such as this various agents have been proven to be effective against mycoplasma (*Table 7*).

Table 7. Antibiotics commonly used in the elimination of mycoplasma in cell cultures

Agent	Supplier	Effective concentration
Ciprofloxacin	Bayer	20 mg l^{-1}
MRA (Mycoplasma Removal Agent)	ICN-Flow	0.5 mg l^{-1}
Novobiocin	Sigma	44 mg l^{-1}
BM Cyclin (2 antibiotics)	Boehringer Mannheim	A (pleuromutilin derivative) 10 mg l^{-1}; B (tetracycline derivative) 5 mg l^{-1}

Protocol 9. Elimination of mycoplasma contamination

Equipment and reagents

- Antibiotic (*Table 7*)
- Materials for subculture (*Protocols 1, 2, and 3*)
- Materials for mycoplasma detection assay

Method

1. Culture cells in the presence of the chosen antibiotic(s) for a period of 10–14 days, during which time most cultures will be passaged approximately four times. Each passage should be performed at the highest dilution of antibiotic that the cell will tolerate, following the manufacturers guidelines.
2. Test the culture for the presence of mycoplasma. If mycoplasma are still detectable it is unlikely that this antibiotic will be successful, and an alternative should be tried on a fresh batch of cells.
3. If the mycoplasma test gives a negative result, then the cells should be cultured in antibiotic-free medium for a period required to conduct ten passages. Testing should be carried out at every passage to monitor treatment success, since mycoplasma may persist at low levels immediately after antibiotic treatment. If the culture is still free of mycoplasma after ten passages in antibiotic-free medium, the mycoplasma may be considered to have been eradicated and a cryopreserved bank of mycoplasma-free cells should be prepared immediately.

8.3 Bacteria and fungi

8.3.1 Sources of contamination

The laboratory environment can provide a rich source of bacterial and fungal contamination, often focused in damp areas such as sinks, waterbaths, humidified incubators, and air-conditioning systems. In addition, materials brought into the laboratory, such as cardboard boxes, can harbour contaminants. In view of this, it is important to remove as much cardboard and paperwork from the laboratory environment as possible. Human operators are also a common source of contamination. This can be overcome in part by the routine use of protective clothing such as gloves and gowns, which are changed at regular intervals.

8.3.2 Detection of bacteria and fungi

For bacterial and fungal contamination, broad-based tests include Gram's stain and cultivation in broths and on agar plates.

Protocol 10. Detection of bacteria and fungi by cultivation

Equipment and reagents

- Thioglycollate broth (Biomerieux)
- Tryptose soya broth (Biomerieux)
- Positive control strains: *Bacillus subtilis*, ATCC 6633; *Candida albicans*, ATCC 10231; *Clostridium sporogenes*, ATCC 11437.

Method

1. Inoculate in duplicate 1.5 ml of cell suspension into 20 ml thioglycollate and tryptose soya broths.
2. Inoculate in duplicate 10 cfu of *Bacillus subtilis* or *Candida albicans* into 20 ml thioglycollate and tryptose soya broths.
3. Inoculate in duplicate 10 cfu *Clostridium sporogenes* into 20 ml thioglycollate broths.
4. Incubate one broth of each pair at 32°C, and one broth at 22°C for 14 days. Incubate *C. sporogenes* at 32°C only.
5. Visually assess for the presence of bacterial growth, as indicated by turbid broth.

8.3.3 Elimination of contamination

In the event of cultures becoming infected with bacteria and fungi, the best course of action is to discard the cultures and, following extensive decontamination of the tissue culture cabinets and work surfaces, resuscitate 'clean' cell stocks. However, in the case of irreplaceable stocks this may not be practicable. In cases such as this there is a wide range of antibiotics which have been proven to be effective against bacteria and fungi (*Table 8*). The choice of antibiotic used will be dependent upon the Gram positive/negative status of the contaminating organism(s). However, broad-spectrum antibiotics active against both Gram-positive and Gram-negative organisms, such as streptomycin sulfate or kanamycin, negate the need to identify the organism. A common approach is to use a combination of penicillin and streptomycin sulfate, which can be purchased as a cocktail from most suppliers of tissue culture reagents. The method is essentially the same as for the elimination of mycoplasma (*Protocol 9*). Cell cultures should be cultured in the presence of antibiotic for a period of 10–14 days, or four passages, after which time the cultures should be maintained in antibiotic-free medium for four passages. If the culture is still free of bacteria after this time, the contamination can be considered to be eradicated, and a bank of contaminant-free cells should be prepared.

Table 8. Antibiotics commonly used in the elimination of bacterial and fungal contamination in cell cultures

Antibiotic	Working concentration ($mg\ l^{-1}$)	Range of activity
Amphotericin	2.5	Yeasts and other fungi
Ampicillin	2.5	Gram positive and Gram negative bacteria
Gentamicin	50	Gram positive and Gram negative bacteria
Kanamycin	100	Gram positive and Gram negative bacteria, yeasts
Neomycin	50	Gram positive and Gram negative bacteria
Nystatin	50	Yeasts and other fungi
Polymixin B	50	Gram negative bacteria
Streptomycin sulfate	100	Gram positive and Gram negative bacteria
Tetracycline	10	Gram positive and Gram negative bacteria
Penicillin G	100 000 $U\ l^{-1}$	Gram positive bacteria

8.4 Viruses

8.4.1 Sources of contaminants

In theory, any material derived from an animal source may be contaminated with viruses. In practice, a risk assessment of the likelihood of viral contamination should be carried out, taking into account the source of material (e.g. patient history, current or recent viral infections) and the potential sources of exogenous virus to which the cell line may have been exposed. These may include animal serum (bovine serum is a rich source of bovine viral diarrhoea viruses), trypsin (porcine parvovirus can be a contaminant of porcine trypsin), or other contaminated cell lines. Indeed, some cell lines are constructed with assistance from viruses or viral components (e.g. EBV transformation). In addition, some cell lines carry endogenous viruses which may be fully expressed as particles (which may or may not be infectious), for example, many murine hybridomas express retroviruses, and CHO cells express non-infectious retroviruses (14).

8.4.2 Methods of detection

Both broad and specific techniques may be used for detection of virus contaminants (7):

- Microscopic examination for cytopathic effects and giant cells, which may be indicative of viral infections.
- Cell lines and freeze-thaw extracts can be co-cultivated with a range of potential host cells (e.g. Vero, MRC-5, 3T3, primary cells or a cell line of the same species of origin as the material under test) for a period of 21 days, including blind passage. Cells are examined microscopically for cytopathic effects or the formation of giant cells throughout the incubation period. At

the end of the incubation period monolayers can be flooded with a red blood cell suspension to reveal the presence of haemadsorbing viruses. In addition, monolayers can be examined by indirect immunofluorescence with a range of virus-specific antibodies.

- Transmission election microscopy of cell sections can be used to visualize virus particles (Chapter 5).
- Reverse transcriptase (RT) assays can be used to detect retroviruses. Assay conditions should allow the detection of both magnesium and manganese ion-dependent RT activity.
- A variety of species of animals, of different ages, can be inoculated with cell culture extracts as a broad screen for adventitious agents.
- The mouse antibody-production (MAP) test involves inoculation of murine cell culture materials into mice, and subsequent examination of serum for the presence of antibodies to a range of murine viruses (*Table 9*). There are also similar hamster (HAP) and rat (RAP) tests.
- Polymerase chain reaction and other specific antibody- or nucleic acid-based assays can be used for detection of known potential virus contaminants.
- Retrovirus infectivity assays involve the co-cultivation of cell culture materials with cell lines susceptible to a broad range of retroviruses (e.g. *Mus dunni* cells (15)), followed by a sensitive detection assay (e.g. PG4S$^+$L$^-$ assay for Mink Cell Focus-forming murine retroviruses, and an IF assay with appropriate antibodies for ecotropic murine retroviruses.

8.4.3 Viruses of particular concern

Table 10 lists viruses which are predominantly of concern as product contaminants, either for human medical or veterinary uses.

Table 9. Viruses detected by the mouse antibody production test

Ectromelia
EDIM
GD VII virus
Hantaan virus
Lymphocytic choriomeningitis virus
Lactate dehydrogenase-elevating virus
Minute virus of mice
Mouse adenovirus
Mouse encephalomyelitis
Mouse hepatitis
Mouse salivary gland (murine CMV)
Pneumonia virus of mice
Polyoma
Reovirus type 3
Sendai
Thymic virus

Table 10. Viruses of particular concern

Species of origin	Viruses of concern
Human	Epstein–Barr virus, cytomegalovirus, hepatitis B virus, hepatitis C virus, human herpes virus 6, human immunodeficiency virus, human T-cell lymphotropic virus.
Non-human	Herpes virus (saimiri and SA-8), simian cytomegalovirus, encephalomyocarditis virus, simian haemorrhagic fever virus, varicella virus of simians, adenovirus, SV-40, monkeypox, rubeola, Ebola virus, foamy virus, SIV, STLV
Murine	See Table 9: MAP test
Bovine, Caprine, Ovine	Bluetongue virus, bovine adenovirus, bovine parvovirus, bovine respiratory syncytial virus, bovine viral diarrhoea virus, reovirus, rabies virus
Canine	Canine coronavirus, canine distemper virus, canine parvovirus.
Equine	Equine herpesvirus, equine viral arteritis virus
Feline	Feline infectious peritonitis virus, feline panleucopaenia virus
Porcine	Porcine haemagglutinating encephalitis virus, porcine adenovirus, porcine parvovirus, transmissible gastroenteritis virus.

8.4.4 Dealing with contamination

Some viral contaminants are a risk to health, and the only course of action is to destroy the cell line. An example of this situation is contamination of a murine hybridoma cell line with lymphocytic choriomeningitis virus, reovirus, Sendai, or Hantaan virus. However, some viral contaminants can be removed or inactivated by downstream processing of cell culture products (6, 7). With the common production of murine retroviruses by murine hybridoma lines, the development of robust virus removal or inactivation techniques (e.g. low pH, heat, solvent/detergent treatments, filtration, etc.) has enabled safe use of monoclonal antibodies in humans (16). Such virus removal or inactivation methods must be developed and validated with the contaminant virus if possible, or a model virus which is similar in size, nucleic acid, presence of envelope, etc.

8.5 Isoenzyme analysis

Isoenzymes are enzymes, present in most types of animal cells, which catalyse the same chemical reactions but differ in their molecular structure, resulting in different electrophoretic mobilities. These differences can be observed in different species, in different tissues of the same species, or cells of the same tissue from different individuals of the same species. The main causes for this heterogeneity are:

- Multiple loci coding for a different polypeptide chain with the same enzyme activity.

- Multiple allelism where a single locus codes for a specific enzyme, and differences in the sequence can occur between individuals of the same species.
- Post-translational modifications of the enzyme structure.

Analysis of these enzymes can identify the likely species of origin, as the structural configurations of the same enzyme from different species can be differentiated by their unique electrophoretic mobilities. However, to make an accurate identification it is necessary to use at least two or three different enzymes such as nucleoside phosphorylase, lactate dehydrogenase, and glucose-6-phosphate dehydrogenase. The last enzyme can differentiate human cell lines of caucasian and non-caucasian origin, and thus represents a useful means of detecting HeLa cell contamination. A variety of electrophoresis matrices can be used, e.g. starch gel, polyacrylamide gel, 'cellogel', and agarose gel. A method using pre-formed agarose gels in a kit (Authentikit System, Innovative Laboratories, Marshfield, MA) is readily available. This system gives rapid results (3 h), requires very small amounts of sample (1–2 μl) and provides permanent records in the form of dried gels.

8.6 Karyotyping

Karyotyping also enables identification of species of origin, based on the study of the number and morphology of chromosomes present in the cells. To visualize the chromosomes, actively growing cell cultures are arrested at metaphase by the addition of mitotic inhibitors such as colcemid. The cultures are then fixed, and chromosome spreads prepared and examined by light microscopy. The range of chromosome numbers and frequency are recorded and the median number calculated. This is then compared to the published karyotype (2n) for a given species, and the species of origin is then confirmed. The presence of specific marker chromosomes may also aid the confirmation of species. Karyotyping provides a quick means of species identification. However, interpretation requires considerable training and experience.

8.7 DNA fingerprinting

DNA fingerprinting visualizes polymorphic loci distributed throughout the genome on different chromosomes by allowing hybridization between a nucleic acid probe and a large number of repetitive DNA families at the polymorphic loci. These loci are called variable number tandem repeats, and vary in number between individuals. Each individual has a unique set of repeated DNA sequences, which are inherited in Mendelian manner. These patterns can be visualized and identified in a wide range of species by a standard method of DNA fingerprinting, which involves hybridization of *Hinf I* digested Southern blots of genomic DNA with the Jeffreys' probe 33.15 (13), and visualization of the bound probe by autoradiography, detecting either a chemiluminescent or [^{32}P] label. The main uses of DNA fingerprinting are:

- To gain information about genetic stability by assuring consistency between cell stocks produced at different times
- To identify cell lines accurately from a wide range of species
- To demonstrate cross-contamination between cell lines

DNA fingerprinting is a powerful cell characterization method, and is unique in its ability to identify simultaneously cell cultures from many species, and exclude cross-contamination from the same range of cell types.

9. Transportation

The exchange of biological material is now commonplace. However, adequate safeguards must be implemented so that the material in transit does not present any hazard to personnel or the environment. Regulations have been developed and published, with input from organizations including the World Health Organization (WHO), UN Committee of Experts on the Transport of Dangerous Goods, and the International Air Transport Association (IATA). A primary condition of exchange of material by air freight or by post (including air mail) is that material is exchanged only between officially recognized laboratories. In addition, material import and export restrictions or quarantine regulations should be observed. Biological material is classified according to its risk to the individual and risk to the community:

- Group I/Class I – low individual risk and low community risk.
- Group II/Class II – moderate individual risk and limited community risk.
- Group III/Class III – high individual risk but low community risk.
- Group IV/Class IV – high individual risk and high community risk.

9.1 Shipment of growing cultures

There are several basic points which should be observed:

- Items should be securely packaged so that there is no risk of other items getting trapped in them.
- Fragile objects should be packed in a strong plastic material or strong cardboard filled with polystyrene chips, bubble wrap, or paper.
- Liquids (i.e. tissue culture media) should be enclosed in leakproof containers inside a strong plastic or cardboard box filled with absorbent material (i.e. paper or cotton wool).
- Secondary watertight packaging is required for infectious perishable material with a layer of absorbent paper between the two watertight containers.

- Packages should be appropriately labelled as follows:
    1. Delivery address
    2. Sender's address
    3. Urgent medical supplies
    4. Fragile
    5. Incubate at 37 °C/25 °C on arrival
    6. Infectious substance label for group III and group IV

9.2 Shipment of frozen materials on dry ice (solid carbon dioxide)

The basic requirements for shipping cryopreserved cultures are:

- Packages should be designed and constructed to hold low temperatures and permit the release of carbon dioxide gas to prevent the build-up of pressure.
- The vials should be placed in the dry ice directly, unless they are of class III/class IV type, when they should be placed within a metal container which is then placed in the solid carbon dioxide.
- **Carbon dioxide pellets should NOT be placed within the metal container.**
- Packages should be appropriately labelled with the:
    1. Delivery address
    2. Sender's address
    3. Urgent medical supplies
    4. Dry Ice UN 1845, × kg (note: the net weight of dry ice must be included)
    5. Transfer vials to gas-phase nitrogen on arrival

The shipper must ensure that the material is properly classified, packed, marked, and labelled. For air-freight shipments, the 'Shippers Declaration for Dangerous Goods' must be properly completed and signed. The shipper must also make prior arrangements with the consignee and the shipper.

References

1. Freshney, R. I. (ed.) (1992). *Animal cell culture: a practical approach.* Oxford University Press, Oxford.
2. Davis, J. M. (ed.) (1994). *Basic cell culture: a practical approach.* Oxford University Press, Oxford.
3. Freshney, R. I. (ed.) (1994). *Culture of animal cells* (3rd edn). Wiley-Liss Inc, NY.
4. Doyle, A., and Griffiths, J. B., (1998). *Cell and tissue culture: laboratory procedures.* J. Wiley & Sons, Chichester.
5. Advisory Committee on Dangerous Pathogens (1995). *Categorisation of biological agents according to hazard and categories of containment.* HSE Books. Sudbury.

6. U.S. Food and Drug Administration, Packville, MD 20852, USA. (1997). *Points to consider in the manufacture and testing of monoclonal antibody products for human use.*
7. U.S. Food and Drug Administration, Packville, MD 20852, USA. (1993). *Points to consider in the characterization of cell lines used to produce biologicals.*
8. USDA Animal and Plant Health Inspection Service, Riverdale, MD 20737, USA. (1997). *Animal and animal products – requirements for ingredients of animal origin used for production of biologics.*
9. O'Brien, S. J., Kleiner, G., Olsou, R., and Shannon, J. E. (1977). *Science*, **195**, 1345.
10. Jeffreys, A. J., Wilson, V., and Thein, S.L. (1985). *Nature*, **314**, 67.
11. Stacey, G. N., Bolton, B. J., and Doyle, A. (1992). *Nature*, **357**, 261.
12. Stacey, G. N., Bolton, B. J., Morgan, D., Clark, S.A., and Doyle, A. (1992). *Cytotechnology*, **8**, 13.
13. Racher, A. J., Stacey, G. N., Bolton, B. J., Doyle, A., and Griffiths, J. B. (1994). In *Animal cell technology: products of today, prospects for tomorrow* (ed. R. E. Spier, J. B. Griffiths, and W. Berthold). p. 69.Butterworth-Heinemann, Oxford.
14. Dinowitz, M., Lie, Y. S., Low, M. A., Lazar, R., Fautz, C., Potts, B., Sernatinger, J., and Anderson, K. (1992). In *Developments in biological standardization* (ed. International Association of Biological Standardization). Vol. 76, p. 201. Karger, NY.
15. Lander, M. R., and Chattopadhyay, S. K. (1984). *J. Virol.* **52**, 695.
16. Marcus-Sekura, C. J. (1991). In *Developments in biological standardization*, (ed. International Association of Biological Standardization). Vol. 75, p. 133, Karger, NY.

2

Virus isolation

A.J. CANN and W. IRVING

1. Why culture viruses?

There are two reasons why the difficult, expensive, and potentially hazardous procedure of virus isolation may be undertaken. The first is for the diagnostic identification of agents associated with disease. The other is to enable some subsequent experimental manipulation of the virus to be performed, for example, to examine mechanisms of replication or to determine the effectiveness of potential antiviral strategies *in vitro*. The protocols described in this chapter are derived from those used in a routine diagnostic virology laboratory, where the aim of virus isolation techniques is to cultivate and identify viruses from a variety of different types of potentially infected biological material (*Table 1*). Culture of viruses in large quantities for further experimental study relies on exactly the same principles, that is, provision of a suitable cell substrate, inoculation with material known to contain viable virus, and incubation under appropriate conditions.

More than fifty per cent of all cases seen by general practitioners are said to be due to virus infections. Accurate and rapid diagnosis is therefore essential for successful antiviral treatment. The vast majority of virus infections are

Table 1. Potential viral isolates from clinical material

Clinical specimen	Viruses which may be isolated in routine cell culture
Vesicle fluid or swab	Herpes simplex virus (HSV); varicella-zoster virus
Nasopharyngeal aspirate (NPA) and/or throat swab	Respiratory syncytial virus; influenza A and B viruses; parainfluenza viruses 1-4; adenoviruses; rhinoviruses; enteroviruses; HSV; cytomegalovirus (CMV)
Faeces	Enteroviruses; adenoviruses
Cerebrospinal fluid	Enteroviruses; mumps virus
Urine	CMV; mumps virus; adenoviruses
Conjunctival swab	HSV; adenoviruses
Blood (buffy coat)	CMV

sub-clinical, and occur without any overt symptoms. Moreover, much of the cell and tissue damage which occurs in virus infections, and most virus replication and transmission, happens before any clinical symptoms are apparent. Virus isolation requires a minimum of 48 hours and may take up to 14–21 days to provide a definitive result. For this reason alone, alternative techniques such as serological investigation and detection of viral nucleic acids are preferred options for the diagnostic laboratory (described in Chapter 4) (*Table 2*). However, there are situations where, for various reasons, these methods fail to give satisfactory results for diagnostic purposes. In these instances, virus isolation may still be a valuable clinical aid, and numerous examples of such cases are described in this chapter. The specialist techniques required to culture plant viruses are described in Chapter 9.

From a scientific viewpoint, many of the reasons for virus isolation have been overtaken by molecular techniques such as Southern blotting, Western blotting, and amplification of virus genomes and mRNAs using the polymerase chain reaction (PCR). While this is sufficient for many investigations, these techniques have one fundamental disadvantage compared with isolation of viruses. Examination of isolated virus components (proteins or genomes) only permits inference of expected behaviour by comparison with previously established models, but does not allow the biology of a virus to be investigated. Although there are many pitfalls in attempting to extrapolate the behaviour

Table 2. Comparison of diagnostic methods

Method	Advantages	Disadvantages
Experimental animals	The only method available for some viruses; can be used to study pathogenesis.	Expensive; slow - results in weeks/months/years; complex systems, complex results.
Virus isolation in cell culture:	Can be used to study and quantify most viruses; sensitive (when it works).	Slow – results in days/weeks; technically demanding; does not work for all viruses.
Direct detection (e.g. EM):	Rapid - results in hours; works for viruses which cannot be cultured.	Sensitivity? Specificity? Technically demanding; equipment is expensive.
Serology:	Rapid - results in hours/days; relatively sensitive; can be quantitative; well suited to processing large numbers of samples and to automation.	Antibodies in patients are not necessarily indicative of current virus infection; possibly not as sensitive as virus isolation, not as sensitive as nucleic acid based methods.
Molecular techniques:	Rapid – results in hours/days; sensitivity can be adjusted to suit the application; may be technically demanding, but this is being overcome by the use of commercially available kits.	Expensive!

of laboratory isolates of virus *in vitro* to any situation which may pertain *in vivo*, the only way novel mechanisms of virus replication can be discovered is by studying virus replication, and this requires the isolation of viruses in a suitable host system which supports their replication. The rest of this chapter attempts to illustrate methods by which this can be achieved.

2. Viruses not amenable to cultivation

The earliest method for detecting viruses (other than clinical observations of infection) was by the use of experimental animals. Some very important advances were made in this way, for example, Walter Reed's use of mice to isolate yellow fever virus during the building of the Panama Canal in 1900. Although a detailed discussion of experimental infection is beyond the scope of this chapter, animal host systems still have their uses in virology:

- To study viruses which cannot be propagated *in vitro*
- To study the pathogenesis of virus infections
- To test vaccine safety (Chapter 6).

But they are increasingly being discarded for the following reasons:

- Breeding and maintenance of animals infected with pathogenic viruses is time-consuming and expensive.
- Whole animals are complex systems, in which it is sometimes difficult to discern individual events.
- Results obtained are not always reproducible, due to host variation.
- Unnecessary or wasteful use of experimental animals is increasingly regarded as morally objectionable.
- These methods are inferior in some ways to molecular techniques such as PCR.

Unfortunately, in spite of many decades of effort, there are still a considerable number of important human pathogens which will not replicate in any readily obtainable *in vitro* system. Hepatitis B virus (HBV) lacks any non-primate animal hosts, and most knowledge of the biology of this virus is based on the closely related woodchuck and duck Hepadnaviruses (1). Terminally differentiated human hepatocytes support limited HBV replication *in vitro*. Several well-established human hepatoma cell lines such as HepG2 will support intracellular replication and the release of viral particles if cloned DNA is introduced into the cell by transfection, but the only cell types to date reported to be efficiently infected *in vitro* are human embryonal cells. For all practical purposes, this means that the full replication cycle of the virus, and specifically the critical first interactions with the host cell which appear to determine the restricted cell tropism of HBV, cannot be studied *in vitro*. As with HBV, there is as yet no readily reproducible system for *in vitro* replication of hepatitis C virus (HCV) and no non-primate animal host. There are now several published reports of molecular clones of HCV which give rise to infection when injected into chimpanzees. However, these clones do not give rise to efficient infection when transfected into a range of cells *in vitro*, although there is evidence that very low levels of virus replication may occur after transfection. Hepatitis D virus (HDV) is a unique agent with certain features reminiscent of a plant viroid and others which resemble a satellite RNA. This agent is entirely dependent on HBV for replication and so cannot be studied other than in systems where HBV will replicate, or by using molecular biological techniques or direct examination of infected clinical samples.

Unlike the hepatitis viruses, human papillomaviruses (HPV) have a wide cell tropism in terms of binding and attachment, but nevertheless fail to replicate in any monolayer cell culture. This is probably the result of their requirement for full epithelial differentiation, which is required for late replication functions leading to the production of virus particles. As with HBV, much of the knowledge of their biology has come from transfection of HPV DNA into various cell types, molecular analysis of infected clinical material, and from a closely related animal virus which can be propagated *in vitro*, bovine papillomavirus (BPV). Small amounts of infectious virus are produced when fragments of HPV-exposed susceptible tissue (cervical epithelium or human foreskin) are inserted under the skin of athymic nude mice, or virus-exposed keratinocytes are inserted into SCID mice (2).

In contrast to HPV, human parvovirus B19 has a restricted host range, due to the distribution of its receptor (blood group P antigen) on only a few cell types such as bone marrow erythrocyte precursors. It can be propagated in primary bone marrow, peripheral blood, or fetal liver cultures, and in a few leukaemic cell lines *in vitro* (3). While most picornaviruses can replicate in cell culture, a number are difficult or impossible to isolate. Among these commonly encountered viruses, many clinical rhinovirus isolates and hepatitis A virus are fastidious, and particular isolates will only replicate in certain cell or organ cultures. Several Coxsackie A viruses (notably A1, A19 and A22) remain impossible to culture, and can only be isolated by inoculation of newborn mice. Thus a single culture system capable of detecting all viruses which might be encountered is not available. Diagnosis of infection with these viruses must of necessity rely on approaches other than cell culture.

Infection of a cell by a virus is a very specific process – not all viruses are able to infect and replicate in all cells. Thus, a laboratory which intends to isolate a range of viruses from clinical material must have available a corresponding variety of cell substrates. In practice, most laboratories will run only two or three different cell types, each of which will support the replication of as broad a range of viruses as possible. The most commonly used cell cultures are listed in *Table 3*, which also indicates which viruses may replicate in each. The techniques involved in propagation of these cell lines are discussed in Chapter 1.

Although this table includes many of the common viruses encountered in

Table 3. Commonly used cell cultures

Type of culture	Viruses capable of replication
(a) Primary/secondary cultures	
Monkey kidney cells	Influenza viruses, parainfluenza viruses, enteroviruses, mumps virus
(b) Semi-continuous cell lines	
Human embryonic fibroblasts	Herpes simplex viruses (HSV), varicella zoster virus, cytomegalovirus, enteroviruses, adenoviruses, rhinoviruses
(c) Continuous cell lines	
Vero cells (derived from monkey kidney)	HSV, mumps virus
HEp-2 cells	Respiratory syncytial virus, adenoviruses
HeLa cells	Adenoviruses

clinical practice, some important viruses (e.g. human immunodeficiency virus (HIV), human hepatitis viruses, the viral causes of gastroenteritis including rotaviruses and enteric adenoviruses, Epstein-Barr virus (EBV/HHV-4)) are missing, as are several others. Isolation of some of these viruses can be achieved by use of particular specialized cell culture techniques. HIV and EBV can be cultured in human peripheral blood lymphocytes, and some of the hepatitis viruses will replicate in primary hepatocyte cultures or hepatoma-derived cell lines. However the difficulties involved in the provision of a regular supply of these more specialized cultures means that isolation of these viruses is usually restricted to reference or research laboratories. However, protocols for the isolation of these viruses are given below:

Protocol 1. Preparation of primary human peripheral blood mononuclear cells (PBMC)

Equipment and reagents

- CO_2 incubator at 37°C
- Benchtop centrifuge with sealed buckets
- Sterile plastic bottles
- Sterile plastic 50 ml tubes
- Sterile plastic 75 cm^2 tissue culture flasks
- Sterile plastic pipettes
- Haemocytometer
- Dextran sedimentation mixture: 15 g dextran (MW 500000), 15 g D-glucose, 4.5 g NaCl; dissolve in 500 ml deionized water, filter through 0.2 μm membrane, store at 4°C.
- Serum-free medium: 1× RPMI 1640, penicillin 50 units ml^{-1}, streptomycin 50 μg ml^{-1}, 10 mM L-glutamine (Sigma, Life Technologies, ICN Flow, etc.)
- Stimulation medium: 1× RPMI 1640, 20% FBS, penicillin 50 units ml^{-1}, streptomycin 50 μg ml^{-1}, 10 mM L-glutamine, rHuIL-2 (5 units ml^{-1}), 10 mg ml^{-1} PHA (Sigma PHA-P).
- Growth medium: as stimulation medium without PHA.
- Ficoll-Paque (Amersham Pharmacia)

Method

1. Pour fresh whole human blood, anticoagulated by collection in preservative-free heparin, into a sterile plastic bottle. (N.B. Human

Protocol 1. *Continued*

blood should be treated as potentially infectious and appropriate safety precautions must be taken).

2. Add 20 ml dextran sedimentation mixture per 100 ml blood and mix gently.
3. Incubate at room temperature until the red cells sediment (usually 30–60 min).
4. Carefully remove the upper plasma layer and discard the red cells.
5. Spin down lymphocytes from the plasma layer by centrifugation in 50 ml tubes for 10 min at 1000 r.p.m.
6. Discard plasma and gently resuspend cells from each 50 ml tube in 5 ml of serum-free RPMI 1640 medium. Pool samples.
7. Dispense 15 ml room temperature Ficoll-Paque to 4 $\times$ 50 ml tubes (more if necessary). Gently overlay each gradient with lymphocyte preparation.
8. Centrifuge gradients for 20 min at 2250 r.p.m.
9. Carefully harvest cells from interface and resuspend in serum-free RPMI 1640 medium (dilute Ficoll-Paque at least 3$\times$).
10. Pellet cells by centrifugation for 10 min at 1250 r.p.m.
11. Wash cells twice more with serum-free RPMI 1640 medium.
12. Resuspend cells in stimulation medium.
13. Count cells using a haemocytometer and adjust concentration to approximately 2×10^6 ml^{-1}.
14. Incubate at 37 °C in CO_2 incubator for 24–72 h. Cell number should increase 2–5 fold after 72 h.
15. Count cells and resuspend in growth medium at $\sim 1 \times 10^6$ ml^{-1}.
16. Cells may also be frozen in liquid nitrogen for future use (Chapter 1). In this case, wash cells thoroughly with growth medium to remove PHA before freezing.

The usual method for HIV isolation involves cocultivation of infected peripheral blood mononuclear cells (PBMC) with uninfected PHA-stimulated PBMC (*Protocol 1*). More than 95 per cent of cultures from HIV-seropositive patients produce detectable HIV infection by this method. Anticoagulated peripheral blood is the usual clinical specimen for HIV isolation, but the methods given in *Protocol 2* may also be used for other specimens, possibly with lower probability of virus isolation.

Protocol 2. Isolation of human immunodeficiency virus

Equipment and reagents

- CO_2 incubator at 37 °C
- Benchtop centrifuge with sealed buckets
- Sterile plastic bottles
- Sterile plastic 15 ml and 50 ml tubes
- Sterile plastic 25 cm^2 tissue culture flasks
- Serum-free medium: 1 × RPMI 1640, penicillin 50 units ml^{-1}, streptomycin 50 μg ml^{-1}, 10 mM L-glutamine (Sigma, Life Technologies, ICN Flow, etc.)
- Sterile plastic pipettes
- Haemocytometer
- Growth medium: 1 × RPMI 1640, 20% FBS, penicillin 50 units ml^{-1}, streptomycin 50 μg ml^{-1}, 10 mM L-glutamine, rHuIL-2 (5 units ml^{-1}).
- Ficoll-Paque (Amersham Pharmacia)
- PBS or HBSS

Method

1. Anticoagulated (preservative free heparin) peripheral blood (10–20 ml from adults, at least 1–2 ml from infants) should be kept at room temperature and processed as soon as possible (within 24 h).
   N.B. All human blood should be treated as potentially infectious and appropriate safety precautions must be taken.
2. Add an equal volume of PBS or HBSS to whole blood sample and mix gently.
3. Overlay Ficoll-Paque with an equal volume of diluted blood in sterile plastic 15 ml or 50 ml tubes as appropriate.
4. Centrifuge at 800 *g*, 25 °C, for 30 min.
5. Carefully harvest the cells from interface and resuspend in serum-free RPMI 1640 medium (dilute Ficoll-Paque at least 3×).
6. Pellet cells by centrifugation for 10 min at 400 *g*.
7. Wash cells once more with serum-free RPMI 1640 medium.
8. Resuspend cells in 10 ml growth medium.
9. Count cells using a haemocytometer. N.B. Caution must be taken when using a glass haemocytometer with potentially HIV-infected samples.
10. Place 10×10^6 cells and 10×10^6 PHA-stimulated PBMC (*Protocol 1*) in a total volume of 10 ml growth medium in a 25 cm^2 tissue culture flask.
11. Incubate at 37 °C in CO_2 incubator.
12. On days 3, 7, 10, 14, 17, 21: remove 5 ml of supernatant from flask and replace with 5 ml fresh growth medium. Freeze a 1 ml aliquot for testing. Culture may be discontinued after 21 days or if criteria for HIV-positivity are met (e.g. two consecutive p24 antigen assays, of which the second is at least four times greater than the first, *or* two consecutive p24 antigen assays that are out of range (O.D >2)).
13. If culture is negative after 21 days, check an aliquot of cells for viability to detect false negatives due to cytotoxicity.

Epstein-Barr virus (EBV/HHV-4) has a dual cell tropism for human B lymphocytes, in which a non-productive infection occurs, and for epithelial cells, which undergo productive infection. There is no suitable animal host for EBV, and infection of primary epithelial cells is not routinely performed, but the replication/latency of this virus has been studied extensively in transformed human B-cell lines, and this technique has proved to be immensely useful for a number of situations where permanently transformed human B cells are required (e.g. human monoclonal antibody production). Laboratory diagnosis of acute disease due to EBV infection commonly depends on testing for the immune response to EBV by heterophile agglutination, ELISA, and immunofluorescence assays. However, the immune response during acute infection may be significantly delayed, especially in children and immunocompromized patients. It is now possible to detect EBV DNA directly by PCR and it has been suggested that the detection of EBV DNA in peripheral blood is associated with symptomatic infection, although, as a result of viral latency and persistence within B cells, it is difficult to differentiate between acute infection and latent infection using PCR.

Protocol 3. EBV transformation of B lymphocytes

Equipment and reagents

- CO_2 incubator at 37 °C
- Benchtop centrifuge with sealed buckets
- Sterile plastic bottles
- Sterile plastic 50 ml tubes
- Sterile plastic multiwell tissue culture plates
- Sterile plastic pipettes
- Haemocytometer
- Ficoll-Paque (Amersham Pharmacia)
- Serum-free medium: 1 × RPMI 1640, penicillin 50 units ml^{-1}, streptomycin 50 μg ml^{-1}, 10 mM L-glutamine (Sigma, Life Technologies, ICN Flow, etc.)
- Growth medium: 1 × RPMI 1640, 20% FBS, penicillin 50 units ml^{-1}, streptomycin 50 μg ml^{-1}, 10 mM L-glutamine
- Sterile 0.45 μm membrane filters (Millipore)
- Cyclosporin A (Sandoz)

Method

1. Dilute 10–20 ml of heparinized venous blood from a healthy adult volunteer with an equal volume of serum-free RPMI 1640 medium. Blood should be kept at room temperature and processed as soon as possible.
   N.B. Human blood should be treated as potentially infectious and appropriate safety precautions must be taken.
2. Overlay Ficoll-Paque with an equal volume of diluted blood in sterile plastic 50 ml tubes.
3. Carefully harvest the cells from interface and resuspend in serum-free RPMI 1640 medium (dilute Ficoll-Paque at least 3×).
4. Pellet cells by centrifugation for 10 min at 1250 r.p.m.
5. Wash cells once more with serum-free RPMI 1640 medium.

6. Count cells (PBMC) using a haemocytometer and adjust concentration to 2×10^6 ml^{-1} in growth medium.
7. Prepare EBV-containing supernatant as follows:
   (a) Cultivate EBV-producer cells at a density of approximately 1×10^6 ml^{-1} in appropriate medium.
   (b) Pellet cells by centrifugation for 10 min at 1250 r.p.m.
   (c) Pass the supernatant twice through sterile 0.45 μm membrane filters to remove all producer cells.
   (d) Dilute the filtrate with an equal volume of growth medium immediately before use.
8. Mix 0.5 ml PBMC (1×10^6 cells) with 0.5 ml freshly diluted EBV-containing supernatant in a multiwell tissue culture plate. Incubate at 37°C in a CO_2 incubator.
9. After 24 h incubation, carefully remove 0.5 ml of the supernatant in each well and replace with 0.5 ml growth medium containing 2 μg ml^{-1} Cyclosporin A (final concentration 1 μg ml^{-1}).
10. Each week, carefully remove 0.5 ml of the supernatant in each culture and replace with 0.5 ml growth medium containing 2 μg ml^{-1} Cyclosporin A (final concentration 1 μg ml^{-1}).
11. When there are signs of cell proliferation, subcultivation of the starting cultures can be carried out (usually 2–3 weeks after infection).
12. Successful transformation is indicated by blastogenesis and the formation of cell aggregates of proliferating lymphoblasts.

3. Treatment of clinical specimens for virus isolation

There are no restrictions on the nature of clinical material that can be inoculated into cell culture for virus isolation. This section and *Protocols 4–8* outline how clinical specimens should be prepared for inoculation into cell culture tubes. The tubes are then incubated under conditions which most resemble those from where the material was obtained. Most cultures are held at 37 °C, but some viruses, most notably those from the upper respiratory tract, will replicate better at a slightly lower temperature, for example, 33 °C, and in the presence of 5% carbon dioxide. Cultures may be slowly rotated on a roller drum, which improves the aeration of the cell sheet, or held stationary.

Fluid samples (e.g. cerebrospinal fluid (CSF), vesicle fluid, nasopharyngeal aspirate, bronchoalveolar lavage fluid, urine) can be sent to the laboratory in a sterile container, and added to appropriate tissue culture tubes either neat or after dilution in maintenance medium (*Table 4*):

- CSF requires no further processing for virus isolation. One drop of CSF is inoculated into one tube each of HEp2, Vero, MRC5 cells and two tubes of

MK cells, and incubated at 37 °C. Any remaining specimen may be stored at –80 °C.

- Vesicle fluids should be received as dried fluid on a microscope slide and on a swab in VTM. The swab can be treated as in *Protocol 6*. The slide is primarily for electron microscopy (Chapter 5), but may also be cultured by adding two drops of VTM to the spot on the slide, using a sterile pipette, and mixing by gentle suction. Inoculate one drop into each of one tube each of Vero and MRC5 cells. The slide should be discarded into a 'sharps' container for appropriate disposal, and the tubes incubated at 37 °C.

Samples from patients with suspected viral infections of the respiratory tract may be taken in a number of forms, for example, nasopharyngeal aspirate (NPA), sputum, bronchial/lung aspirates, or bronchoalveolar lavage fluid. These specimens should be sent to the laboratory in either a plastic trap bottle or in a universal bottle. Rapid diagnosis by immunofluorescence may also be requested on these samples.

Table 4. Transport media

Faecal transport medium (FTM)	
HBSS	1000 ml
Sodium bicarbonate solution (4.4% saturated with CO_2)	13.5 ml
Colomycin (colistin) 100 000 units ml^{-1}	2ml
Fungizone (amphotericin B) 5 mg ml^{-1}	2 ml
Gentamicin 4000 units ml^{-1}	2 ml
Penicillin 0.2 mega units ml^{-1}	10 ml
Urine transport medium (UTM)	
Sterile de-ionised water	810 ml
Newborn calf serum	100 ml
MEM (10x)	90 ml
Sodium bicarbonate solution (4.4% saturated with CO_2)	50 ml
Sorbitol (50%)	50 ml
Colomycin (colistin) 100 000 units ml^{-1}	1 ml
Fungizone (amphotericin B) 5mg ml^{-1}	1 ml
Gentamicin 4000 units ml^{-1}	1 ml
Penicillin 0.2 mega units ml^{-1}	5 ml
Virus transport medium (VTM)	
HBSS	500 ml
BSA (7.5%)	13.5 ml
Sodium bicarbonate solution (4.4% saturated with CO_2)	4 ml
Colomycin (colistin) 100 000 units ml^{-1}	0.5 ml
Fungizone (amphotericin B) 5mg ml^{-1}	0.5 ml
Gentamicin 4000 units ml^{-1}	0.5 ml
Penicillin 0.2 mega units ml^{-1}	1.25 ml

Protocol 4. Processing of samples from the respiratory tract for virus isolation

Equipment and reagents

- Sterile bijou bottles
- CO_2 incubators at 33 °C and 37 °C
- Vortex mixer
- MK, MRC5, HEp-2, and Vero cell cultures
- Sterile plastic pipettes
- PBS
- UTM (*Table 4*)
- VTM (*Table 4*)

Method

1. If the NPA specimen is particularly mucoid, make a saline solution by adding 2–3 ml sterile PBS and vortexing thoroughly. Sputum should be treated as for an NPA. If particularly mucoid, sputum can be diluted 1:2 in sterile PBS, vortexed vigorously, centrifuged at 4000 r.p.m. for 20 min, and the pellet resuspended in PBS. 2–3 ml of bronchial and alveolar aspirates and lavages should be mixed with an equal volume of UTM.
2. Divide the suspension between two sterile bijou bottles. Retain one for immunofluorescence. To the other add an equal amount of virus transport medium, mix and use for virus isolation.
3. Inoculate each of the tubes of a respiratory set with four drops of suspension.
4. Incubate MK and MRC5 cell cultures at 33 °C, and the HEp-2 and Vero cell cultures at 37 °C.

Protocol 5. Processing of urine for virus isolation

Equipment and reagents

- 80 °C freezer
- CO_2 incubator at 37 °C
- MK, MRC5, HEp-2, and Vero cell cultures
- Sterile plastic pipettes
- PBS
- UTM (*Table 4*)

Method

1. Urine samples should arrive in UTM. If not, mix with an equal volume of UTM.
2. Inoculate five drops into one tube each of MK, HEp2, Vero, and MRC5 cells.
3. Incubate at 37 °C.
4. Save an aliquot for storage at –80 °C.

Protocol 6. Processing of swabs for virus isolation

Equipment and reagents

- 80 °C freezer
- CO_2 incubator at 37 °C
- Vortex mixer
- Appropriate cell cultures for target viruses
- Sterile plastic pipettes

Method

1. Swabs (e.g. conjunctival, throat, base of ulcer, cervical, rectal) should arrive broken off into VTM (isotonic fluid plus antibiotics to inhibit bacterial overgrowth). Dry swabs or those in bacteriological transport medium are unacceptable, and repeat specimens, in the correct medium, should be requested.
2. Ensure the container top is tight. Vortex mix for 15 s to dislodge material on swab into the transport medium.
3. Inoculate each of the appropriate tubes with approximately six drops of the transport medium using a disposable sterile plastic Pasteur pipette.
4. Incubate tubes at 37 °C.
5. Store remainder of specimen at –80 °C.

Protocol 7. Processing of tissue for virus isolation

Equipment and reagents

- 4 °C refrigerator
- CO_2 incubator at 37 °C
- Benchtop centrifuge
- Stomacher machine or Griffiths tubes (e.g. Seward Ltd. London, UK)
- Appropriate cell cultures for target viruses
- Sterile plastic pipettes
- Sterile universal tubes
- Sterile bijou bottles
- FTM

Method

1. Fresh tissue (e.g. brain, bowel), both biopsy and post-mortem, should be received in dry, sterile containers, for example, plastic universals.
2. Place tissue in a sterile dish.
3. Using scalpel and forceps, remove a representative piece of tissue, approximately 1 cm cube, and place in a plastic 'stomacher' bag. Add 10 ml FTM, fold bag, and place inside a second bag.
4. Place in 'stomacher' machine and homogenize for 1–2 min. Alternatively, tissue can be ground up in a Griffiths tube.
5. Decant contents of inner bag into a universal tube and centrifuge at 4000 r.p.m. for 20 min.

6. Inoculate four drops of supernatant into each of the appropriate tubes and incubate at 37 °C.
7. Pipette approximately 4 ml of supernatant into a bijou bottle and store at –80 °C.
8. The remainder of the specimen, including deposit, may be stored at 4 °C for possible electron microscopy.

Protocol 8. Processing of stool samples for virus isolation

Equipment and reagents

- 4 °C refrigerator
- 80 °C freezer
- CO_2 incubator at 37 °C
- Benchtop centrifuge
- Vortex mixer
- MK, HEp2, Vero, and MRC5 cell cultures
- Sterile plastic pipettes
- Sterile bijou bottles
- FTM (*Table 4*)

Method

1. Add approximately 1 g of stool to 10 ml FTM, and vortex to emulsify (ensure the cap is firmly screwed down).
2. Centrifuge at 4000 r.p.m. for 20 min in benchtop centrifuge.
3. Inoculate one tube each of MK, HEp2, Vero, and MRC5 cells with two drops of supernatant and incubate at 37 °C.
4. Pipette approximately 4 ml supernatant into a bijou bottle and store at –80 °C. The remainder of the specimen (including deposit) should be stored at 4 °C for electron microscopy.

4. Identification of virus replication

The presence of a virus replicating within a cell may have profound effects on the function of the cell. Most virus-infected cells will undergo morphological changes, for example, swelling due to alteration of membrane permeabilities, shrinkage due to cell death, and formation of giant multinucleate cells, which are visible under the light microscope, due to the presence of a fusion protein encoded by the virus. Collectively, these changes are referred to as cytopathic effects (CPE). Thus, observation of an inoculated cell sheet over a period of time for the presence or absence of a CPE is a straightforward screening method for determining whether or not a virus is present in the cell culture. This can be conveniently performed using the low-power lens of a light microscope (*Figure 1*).

The exact appearance of a CPE depends on the nature of the virus present

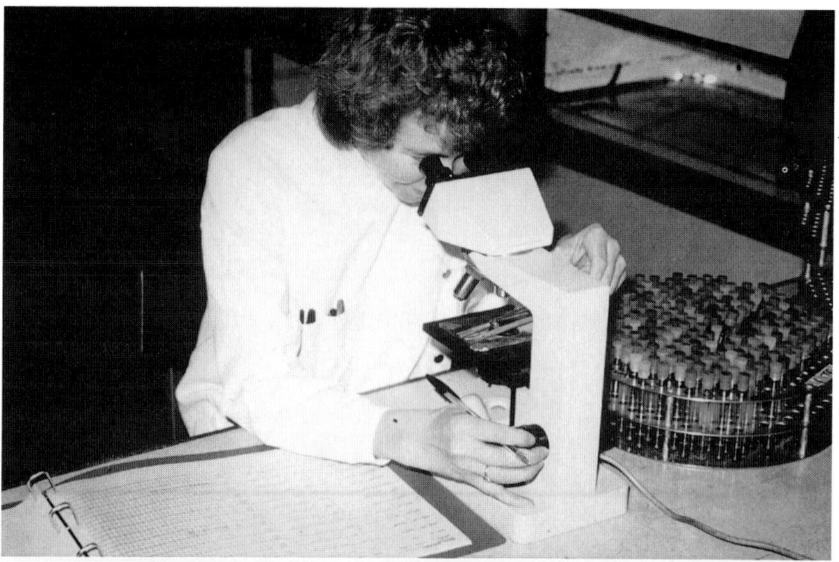

Figure 1. Each tube on the roller (seen on the right of the microscope) is examined under low power for evidence of a cytopathic effect.

and the particular cell substrate in use. Cells may swell or shrink; the effect may be spread over the whole sheet, or be confined to small foci; or it may occur predominantly, if not exclusively, at the edge of the sheet rather than in the middle. The time for development of the CPE may vary from as short as 18 h to as long as 4 weeks. Consideration of all of these factors, plus the clinical details relevant to a particular specimen, will allow a trained virologist to make a definitive diagnosis of which virus is present in a given culture through light microscopy observation of the CPE alone. Thus, a CPE originating from a vesicle swab consisting of ballooning cells, appearing after 24 hours of culture and thereafter spreading rapidly to destroy the whole cell sheet, is highly likely to be due to herpes simplex virus. In contrast, a vesicle swab causing a CPE to appear only after 10 days in culture, in only two or three foci of swollen cells in the centre of the sheet which spread slowly over the next several days, is likely to be due to varicella-zoster virus. Illustrations of CPEs characteristic of specific viruses are shown in *Figure 2*.

5. Passaging of viruses

Cell sheets may deteriorate and therefore exhibit morphological changes due to factors other than infection by virus. Material inoculated onto the sheet may be toxic, a problem especially frequent with faecal samples, or bacterial or fungal contamination may occur, despite the presence of antibiotics and

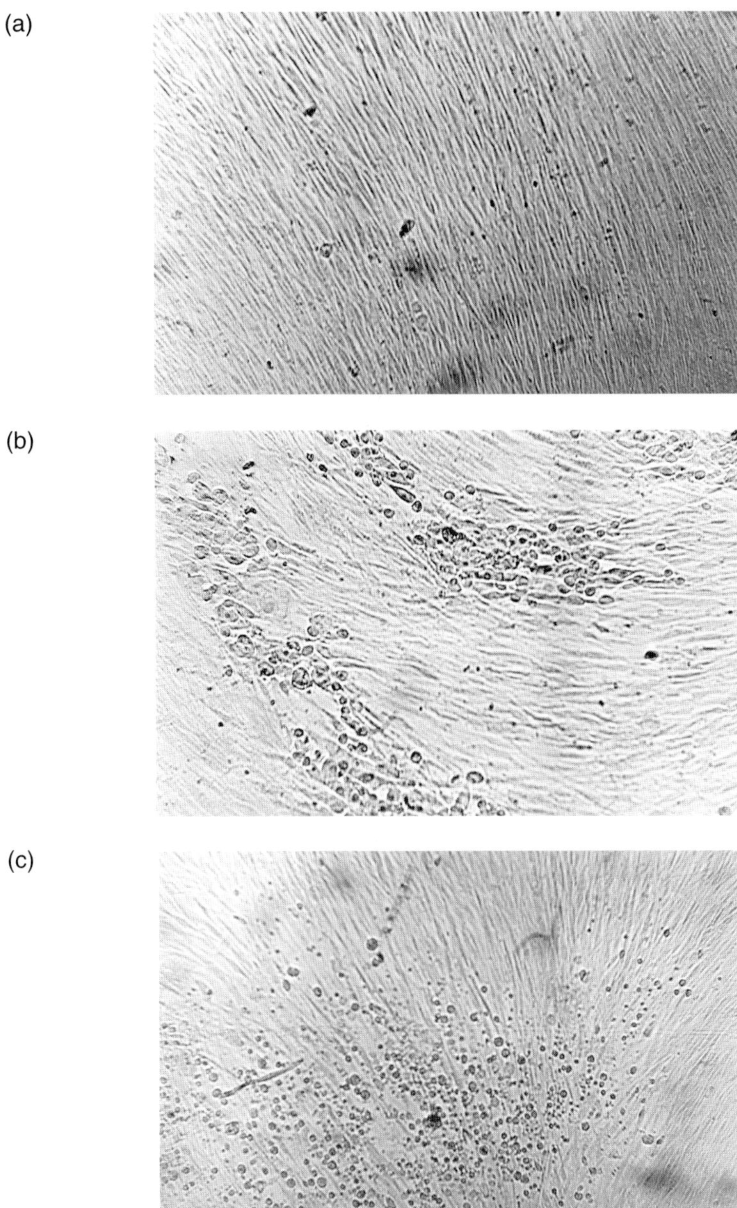

Figure 2. Cytopathic effects (CPE). (a) A normal cell sheet of fibroblasts (MRC5 cells). The cells have grown to confluence, and are spindle-shaped. (b) Some of the cells are rounded up and swollen. This appearance in discrete foci separated by normal-looking cells is typical of the cytopathic effect induced by cytomegalovirus. (c) Several cells all over the cell sheet have rounded up to give a 'flick-drop' appearance. This is the characteristic appearance of a rhinovirus-induced CPE.

antifungal agents in the culture medium. If there is any doubt as to the origins of a CPE, that particular culture should be passaged into fresh cells, which are then incubated in place of the original and examined regularly to see if the effect is reproducible. Using a sterile straw, cells are scraped into the medium, and a small aliquot withdrawn. This is transferred to a fresh tube of the same cell type (and others if necessary). The original tube may be stored at 4 °C for 3 days if required for further reference. If the original CPE arose because of a virus infection, then it should reappear in the passaged tube. As the titre of virus will have increased within the first culture tube, the time to appearance of CPE in the passaged tube should be considerably shortened. Alternatively, if the original CPE was due to non-specific toxic factors, then these should have been diluted during passage, and the subsequent damage to the cell sheet in the passaged tube should appear more slowly and less extensively, if at all.

There are other reasons why it may be necessary to passage viruses. Some viruses fail to produce a visible CPE on initial *in vitro* culture, despite viral replication. However, adaptation of the virus to the particular cell substrate may occur on repeat passaging, resulting in the production of a visible CPE. This process of passaging normal-looking cell cultures in order to encourage the appearance of a visible CPE is known as *blind* passage. The identification of rubella virus replicating in rabbit kidney cells may require several such blind passages. Secondly, passaging of viruses is a useful way of increasing the amount of virus available for further studies, such as neutralization assays (see below).

6. Confirmation of virus isolation

In many instances the type of virus in a cell culture can be determined with reasonable certainty by observation of the characteristics of the CPE present in the cell sheet, as outlined above. However, there are instances where confirmation of virus identity may be required, for example, if the CPE is atypical, does not passage well, or is absent, or alternatively, when more information about the particular strain of virus is required. This can be achieved by use of electron microscopy, antigen detection, acid stability, or neutralization assays.

Enteroviruses and rhinoviruses are both common isolates, producing shrinkage and rounding of MRC5 cells. A CPE arising from an enterovirus usually spreads rapidly throughout the whole cell sheet, whereas rhinoviruses tend to form non-confluent foci. Sometimes, however, this distinction can be blurred, for example, if there is a particularly high titre of rhinovirus, or an atypical enterovirus. Enteroviruses are stable at pH 2, whilst rhinoviruses are acid labile. This difference in viral properties is the basis for a test to differentiate the two virus genera.

Protocol 9. Acid stability test for the differentiation of enteroviruses and rhinoviruses

Equipment and reagents

- Medium A (Eagles MEM without bicarbonate, pH 2) (Sigma, Life Technologies, ICN Flow, etc.)
- Medium B (medium A adjusted to pH 7 with bicarbonate) (Sigma, Life Technologies, ICN Flow, etc.)
- CO_2 incubator at 37 °C
- MRC5 roller tube cultures
- Sterile plastic pipettes
- Sterile bijou bottles

Method

1. Add 0.1 ml volume of infected cell culture fluid to two bijou bottles, one containing 0.9 ml medium A, and the other containing 0.9 ml medium B.
2. Incubate at room temperature for 2 h.
3. Titrate from vials A and B in tenfold dilutions in medium B from 10^{-1} to 10^{-4}.
4. Inoculate 0.2 ml of each dilution into human embryonic fibroblast (MRC5) cells. Set up a virus control (0.2 ml from vial B) and a cell control (uninoculated).
5. Incubate tubes rolling at 33 °C, and examine for CPE at 3 and 5 days. A virus able to replicate derived from both A and B bijoux is an enterovirus, whilst a rhinovirus will only replicate in tubes inoculated from bijou B.

The process of cell culture acts as a means of increasing viral titre to a sufficient level to allow visualization by electron microscopy (EM) (Chapter 5). Thus, provided there is access to an EM, most problems associated with atypical or unusual CPEs can be resolved rapidly by appropriate EM examination of the cells or culture supernatant.

Neutralization assays are used most commonly to distinguish between the various enteroviruses which replicate in culture, or for serotyping adenoviruses, but can in principle be used to confirm the nature of any virus isolate, provided a suitable antiserum is available (see Chapter 7). The basis of this test is to use a set of specific antisera to neutralize the virus, and thereby prevent infection of new cells. Virus isolated in culture is distributed into a series of tubes, and to each tube a different antiserum is added. The type of cell and the nature of the cytopathic effect should give a reasonable indication as to the virus present. This will determine the range of antisera used in the test (e.g. anti-polio-1, anti-polio-2, and anti-polio-3 to tubes 1, 2, and 3 respectively). Following incubation of virus plus antiserum for one hour, the mixtures are added to individual tissue culture tubes, and the appearance of CPE is monitored over the next few days. The occurrence of a typical CPE in tubes 1 and 3, combined with the absence of a CPE in tube 2, would indicate that the original virus was poliovirus type 2.

Protocol 10. Viral neutralization test

Equipment and reagents

- CO_2 incubator at 37 °C
- MRC5 roller cultures
- Sterile plastic pipettes
- Sterile bijou bottles
- FTM (*Table 4*)
- HBSS
- Appropriate antisera

Method

1. Perform a rough estimate of the tissue culture infectious dose of virus present in the primary isolate tube by inoculating two drops of serial tenfold dilutions of stock virus into appropriate tissue culture tubes. Observe tubes over the next few days for evidence of CPE.
2. Label bijou bottles with the name of the antisera to be used and the laboratory number of the virus isolate. Label one vial as the virus control.
3. Prepare any antisera necessary for the test. Antisera are made up at 10× titre in FTM (e.g. 0.2 ml vial serum with a titre of 1:200 will require a final volume of 4 ml). Check that the antisera are clear and not floccular or contaminated.
4. Add four drops of the appropriate antiserum to the equivalent bijou using separate Pasteur pipettes for each antiserum. Add four drops of HBSS to the virus control vial.
5. Dilute the virus stock to a dilution known to induce an extensive CPE within a convenient time period, e.g. 48–72 h, as determined in step 1 above, in FTM. This equates to around 100 $TCID_{50}$. Add one drop of the diluted virus to each bijou.
6. Mix and incubate the virus plus antisera for 30 min at room temperature.
7. Label new cell culture tubes with specimen number, antiserum, and date of test. Inoculate the entire contents from each bijou into the appropriate tube. Leave the cell control uninoculated.
8. Incubate the tubes at 37 °C. Examine the vials daily for CPE.

Cells infected with a particular virus will express antigens derived from that virus on their surface. These antigens can be detected by staining with antibodies. Thus, the specific identity of a virus replicating in cell culture may be determined by staining the cells with a panel of appropriate monoclonal antibodies, suitably labelled with an immunofluorescent tag. Cells can either be scraped off and sucked out of the tissue culture tubes, and then dried onto a glass slide (*Protocol 11*), or, if it is known in advance that antigen detection will be necessary, the cells can be grown on a flat surface such as a coverslip within the tissue culture tube, which can be retrieved subsequently and stained. This latter form of culture is known as a shell vial. Preparation and staining of shell vials is described as part of the DEAFF (*Protocol 13*). Antigen detection will allow distinction between, say, influenza A and B viruses, or parainfluenza 1, 2, or 3 viruses, which all produce a similar CPE (Chapter 4).

Protocol 11. Preparation of cell culture isolates for immunofluorescence

Equipment and reagents

- Refrigerator at 4 °C
- Incubator at 37 °C
- Multipoint immunofluorescence slides (e.g. ICN Biomedicals)
- Sterile plastic pipettes
- Sterile bijou bottles
- PBS
- Acetone

A. *Preparation*

1. Cell cultures containing haemadsorbing viruses or cytopathic viruses for typing (e.g. differentiating herpes simplex type 1 and 2 viruses) or isolation confirmation may be processed as follows (N.B. A slide of uninfected control cells should be included and treated as for virus infected cells):
   (a) Haemadsorbing virus infected cells: incubate the tube for 30 min at 37 °C to allow the red blood cells (RBC) to elute. Decant the RBC suspension.
   (b) Other virus infected cells: decant the medium into a sterile bijou and retain for reference purposes (at 4 °C or long-term at –80 °C).
2. Add three drops of PBS to the tube, and scrape the cells into the solution using a sterile wooden applicator stick or a Pasteur pipette.
3. Deliver the suspension to the appropriate number of spots on a degreased labelled multispot slide. Allow to air-dry, which may be aided by gentle heat from a warm plate.
4. Fix by immersion in acetone for 10 min.
5. Stain immediately or store at 4 °C. If the slide cannot be stained within 48 h, store at –80 °C.

B. *Staining*

Most antibodies in present use are commercial monoclonal reagents for use in direct immunofluorescence (IF) tests. The manufacturers tend to

Protocol 11. *Continued*

recommend using excessive volumes of reagent per test. In practice, volumes may be greatly reduced without loss of sensitivity. Different manufacturers may recommend different staining protocols, e.g. varying the incubation/washing temperatures and times, but a compromise protocol is best adopted such that a single slide may be tested against a range of reagents.

B. *Direct immunofluorescence*

1. Apply 10 μl of diluted antibody per spot.
2. Incubate the slide in a moist box at 37 °C for 30 min.
3. Rinse the slide carefully with PBS using a wash bottle. Place the slide in a staining rack and immerse in PBS for 10 min.
4. Rinse in distilled water.
5. Dry the slide (in air, or using a hair-dryer).
6. Mount with coverslip using the manufacturer's mountant.
7. Examine under UV light.

C. *Indirect immunofluorescence*

1. Apply 10 μl of the first indirect immunofluorescence reagent, i.e. specific antibody, to the appropriate spot(s). Apply 10 μl PBS to a conjugate control spot.
2. Incubate the slide in a moist box at 37 °C for 30 min.
3. Rinse the slide carefully with PBS using a wash bottle. Place the slide in a staining rack and immerse in PBS for 15 min.
4. Rinse in distilled water. Air dry.
5. Apply 10 μl of FITC conjugate to each test spot and to the conjugate control spot.
6. Reincubate the slide in a moist box at 37 °C for 30 min.
7. Wash and dry the slide as in steps B3–B5.
8. Mount with coverslip using the manufacturer's mountant.
9. Examine under UV light. Examine the conjugate and uninfected cell controls before the test spots.

7. Adaptations of cell culture

One of the major drawbacks of virus isolation as a means of laboratory diagnosis of infection is the length of time taken for some viruses to produce a visible CPE. The fastest replicating viruses, such as herpes simplex or polio virus, may produce damage to the cell sheet in as little as 24 hours, but some viruses, for example, cytomegalovirus and adenoviruses, may take up to 4 weeks to do so. A variety of modifications of cell culture have therefore been developed, with the aim of speeding up the process of virus isolation. These are described in this section.

(a)

(b)

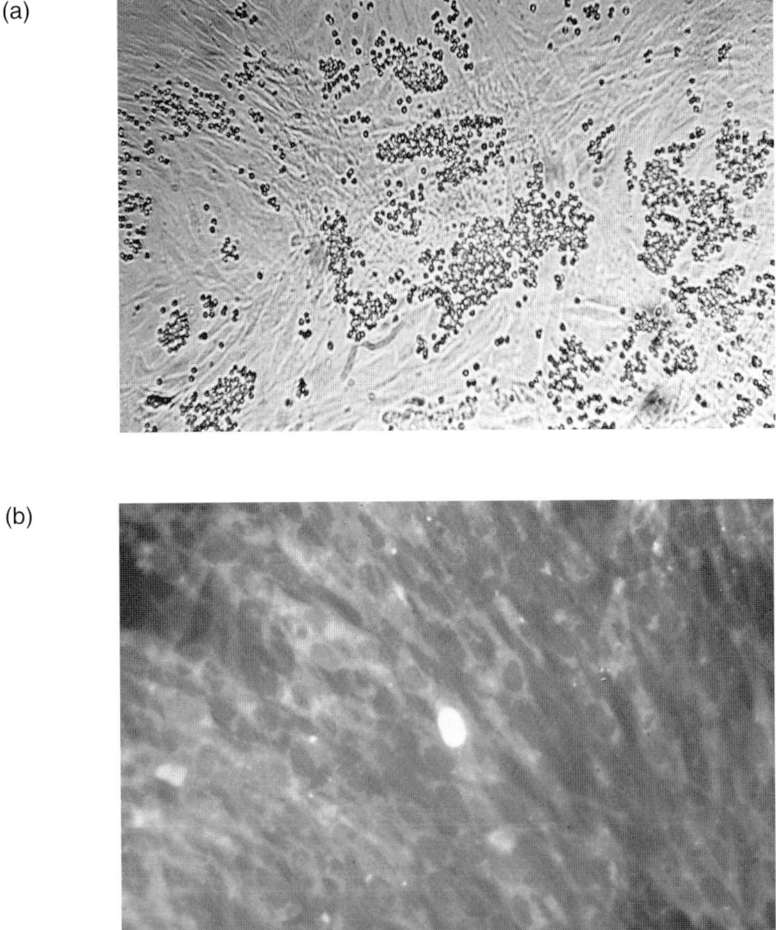

Figure 3. Adaptations of cell culture. (a) Haemadsorption. The cell sheet (primary monkey kidney) is normal, that is, no visible cytopathic effect has yet developed. However, many cells within the sheet are expressing a virus-encoded haemagglutinin on their surface, since the red cells, here seen as small dots overlying the cell sheet, have become attached to the sheet. (b) The DEAFF test. The cell sheet is stained with a fluorescently-labelled anti-CMV monoclonal antibody. A single nucleus is seen shining bright apple-green under UV light, indicating the presence of CMV in this cell. All the cells (seen here counterstained with Evans' blue dye) are morphologically normal. (Reprinted with permission from *The Science of Laboratory Diagnosis*, ISIS Medical Media)

7.1 Haemadsorption test

Many viruses (e.g. influenza and parainfluenza viruses, mumps virus) contain surface glycoproteins known as haemagglutinins (HA), capable of binding red blood cells. As these viruses replicate in cell culture, HA molecules appear on the cell surface. Thus, if red blood cells from appropriate species are added to a cell culture tube in which such a virus is replicating, they will adhere to the cell sheet, a phenomenon known as haemadsorption (*Figure 3a*). The presence of a haemadsorbing virus can therefore be detected several days before a CPE becomes evident. Exact identification of the haemadsorbing virus can be determined by a variety of techniques, including immunofluorescence with monoclonal antibodies (*Protocol 11*).

Most viruses capable of haemadsorption also produce neuraminidase (NA) on their surface. This enzyme may destroy the HA attachment sites on the red cells, and thereby give rise to a false negative result. This potential source of error is eliminated by performing the assay at 4°C, a temperature at which the NA is inactive.

Protocol 12. Haemadsorption test

Equipment and reagents

- Refrigerator at 4°C
- CO_2 incubator at 37°C
- Benchtop centrifuge
- Immunofluorescence microscope
- Multipoint immunofluorescence slides (e.g. ICN Biomedicals)
- Sterile plastic pipettes
- Sterile bijou bottles
- Guinea pig blood (in Alsevers solution)
- PBS
- HBSS
- Acetone

Method

1. Select tubes to be tested and place in rack.
2. Label bijou vials with the specimen number of the tubes to be tested.
3. Prepare a suspension of guinea pig red blood cells by adding 1 ml of stock blood (provided in Alsevers solution) to 20 ml HBSS. Centrifuge at 2000 rpm for 5 min. Decant supernatant and replace with fresh HBSS. Repeat this washing procedure 3 times. From freshly washed cells prepare a 0.5% suspension of cells in HBSS.
4. Decant the cell culture fluids into their corresponding bijoux.
5. Add 0.5 ml red cell suspension to each tube.
6. Place the tubes at 4°C for 30 min. Make sure the cell sheets are facing down and covered by the suspension.
7. Examine the tubes for haemadsorption, one rack at a time to avoid warming up, resulting in NA-mediated erythrocyte elution. HA is best seen by gently rotating the tubes under the microscope. Adsorbed

cells will be seen to be stuck to the cell sheet, whilst non-adsorbed cells will move freely in the culture supernatant (*Figure 3a*).

8. HA-negative tubes may be discarded if they have fulfilled the necessary length of time in culture. Otherwise the cell sheet should be washed once in fresh medium, 1 ml medium added, and the tubes returned to the drums for further incubation.
9. Incubate HA-positive tubes at 37°C for 30 min to elute the red cells. Decant the fluid and carefully wash the cell sheet.
10. Add 0.1 ml PBS and gently scrape off the infected cells. Pipette these cells on to six wells of a multipoint slide. Use similarly prepared uninfected cells and spot onto the remaining six wells.
11. Allow the multispot slide to air dry. Fix in acetone for 10 min. The slide is now ready for identification by immunofluorescence.
12. The original culture fluid (stored in bijou bottles, step 4 above) may be passaged to culture virus up for other tests (e.g. neutralization) if required. Store positive vials at 4°C until identification is complete. Discard the negative bijoux.

7.2 Detection of early antigen fluorescent foci (DEAFF test)

This test allows rapid diagnosis of cytomegalovirus (CMV) infection. It is essentially a shortened isolation technique using immunofluorescence (IF) to detect the presence of virus, rather than wait for the appearance of a cytopathic effect. It involves the inoculation of a specimen onto a monolayer of MRC5 cells growing on a coverslip or other flat surface (a type of culture known as a shell vial), centrifuging at low speed to enhance infectivity, incubating for 24–48 h, and then testing the cells for the presence of a CMV-specific antigen found in the nuclei of infected cells using a monoclonal antibody IF reagent. The antigen to be detected is selected as one which appears early (within 24–48 h) in the replication cycle of the virus (hence 'early antigens').

Protocol 13. DEAFF test

Equipment and reagents

- CO_2 incubator at 37 °C
- Benchtop centrifuge
- Plastic shell vials (e.g. Redhill Surgical, Glasgow, UK)
- Sterile glass coverslips
- Sterile plastic pipettes
- Sterile bijou bottles
- MRC5 cells
- VTM
- PBS
- Fixative: 50:50 acetone:methanol
- Growth medium: 1 × MEM, 10% FBS, penicillin 50 units ml^{-1}, streptomycin 50 $\mu g\ ml^{-1}$, 10 mM L-glutamine.
- Anti-CMV Nuclear Early Antigen (NEN cat #NEA9221). Reagent is provided as freeze-dried powder. Add 1.0 ml sterile distilled water and allow to dissolve. Dispense 25 μl aliquots in plastic serology vials and store at −80°C. Remove vial from −80°C and allow to thaw. Add 0.4 ml sterile PBS and mix. Store at 4 °C once diluted.

Protocol 11. *Continued*

A. *MRC5 shell vial monolayer preparation*

1. Trypsinize MRC5 cells from a confluent 75 cm^2 flask into 10 ml growth medium.
2. Make half of this cell suspension up to 12 ml with growth medium and dispense 1 ml into each of twelve plastic shell vials containing sterile glass coverslips.
3. Shake each vial to remove any air bubbles from under the coverslip and incubate at 37 °C in a CO_2 incubator.
4. The cell sheet should be confluent within 48 h after which they should be changed to maintenance medium. Confluent sheets may be kept for up to 7 days if their medium is changed regularly.

B. *Specimen preparation*

1. Urine should arrive in UTM (equal volumes). Centrifuge at 2000 r.p.m. for 10 min. Remove the supernatant saving 5 ml for routine virus isolation. Resuspend the deposit in approximately one tenth of the initial volume of supernatant, and use this to inoculate the vial. Also inoculate into cell culture tube for routine virus isolation.
2. Bronchial and alveolar aspirates. If mucoid, these should be diluted in PBS. Mix with an equal volume of UTM.
3. Swabs, e.g. throat swabs. These should be received in VTM. Vortex for 10 s and use this medium.
4. Blood. The DEAFF test can be performed on buffy coat cells from a sample of peripheral blood, but rapid diagnosis of CMV infection in a blood sample is better done by the CMV antigenaemia assay (*Protocol 14*).

C. *Shell vial assay – inoculation*

1. Decant medium from confluent MRC5 cell shell vial.
2. Inoculate 0.5 ml specimen into shell vial.
3. Centrifuge the shell vial at 2000 r.p.m. for 1 h at 30 °C.
4. Add 1 ml MEM maintenance medium and incubate at 37 °C in a CO_2 incubator overnight.

D. *Fixation – performed in shell vial*

1. Decant medium from shell vial.
2. Carefully rinse coverslip with 1 ml PBS, decant.
3. Carefully rinse with 1 ml fixative (50:50 acetone:methanol). Decant.

4. Fix in 1 ml fresh fixative for 10 min.
5. Decant fixative.

E. *Staining*

The following method describes an indirect immunofluorescence (IF) assay. Directly conjugated anti-CMV early antigen reagents are available, so this can be performed as a direct IF assay if desired.

1. Remove coverslip and allow to air-dry briefly.
2. Deliver 8 μl of anti-CMV nuclear early antigen reagent onto a clean slide and transfer coverslip, cell-side down, onto the reagent.
3. Incubate at 37 °C in a moist box for 30 min.
4. Wash with PBS for 5 min. Rinse with distilled water.
5. Dry under a hair dryer.
6. Deliver 8 μl anti-mouse FITC conjugate onto a clean slide and place coverslip, cell-side down, onto the conjugate.
7. Incubate at 37 °C in a moist box for 30 min.
8. Wash in PBS as before. Rinse in distilled water, dry with a hair dryer, and mount coverslip cell-side down in glycerol.

For CMV infection, fluorescence is confined to the cell nucleus. Nuclei appear oval or kidney-shaped, and surprisingly large. Nucleoli are often apparent. The staining pattern can vary from small bright nuclear inclusions for extremely early infections to a uniform finely granular staining of the entire nucleus. Cytoplasm and 'negative' nuclei will appear unstained (*Figure 3b*). This technique has been extended to other slow-replicating viruses – for example, eye swabs which may contain adenoviruses can be inoculated exactly as described for the DEAFF test onto shell vials of HEp2 cells. After incubation at 37 °C in a CO_2 incubator for 3 days, the coverslip can be fixed and stained with a FITC-conjugated monoclonal antibody against adenovirus, as described for the DEAFF test. Positive cells are seen singly or occasionally in clumps, with bright apple-green nuclear or whole-cell fluorescence. Negative cells appear a dull red colour.

Demonstration of CMV viraemia (i.e. the presence of replicating CMV in a blood sample) is considered more clinically relevant than detection of virus in urine and/or respiratory secretions. Alternative assays to the DEAFF test have been developed for this purpose, including detection of CMV DNA or RNA in plasma, serum or circulating cells by nucleic acid amplification techniques. At present, many of the latter assays described in the literature are 'in-house', and the lack of standardization of protocols across different laboratories creates difficulties in comparison of results. A test based on the detection of CMV antigen-positive polymorphonuclear leucocytes (neutrophils) is both sensitive (more so than the DEAFF in this setting), and gives quantitative results of clinical significance. The CMV pp65 antigenaemia test involves purification of leucocytes from a fresh (less than 6 h old) specimen of EDTA-anticoagulated blood, preparation of duplicate cytospin samples of the cells (100 000 cells per slide), fixation of the slides in formaldehyde, permeabilization of the cells with TRX-100, and then staining with a mouse monoclonal antibody directed against CMV pp65 tegument protein using an indirect fluorescent technique.

Protocol 14. CMV antigenaemia test

Equipment and reagents

- Benchtop cytocentrifuge
- Incubator at 37 °C
- Sterile plastic 15 ml tubes
- Sterile plastic pipettes
- Haemocytometer
- Histopaque 1119 (Amersham Pharmacia)
- PBS

A. *Preparation of slides:*

1. Carefully layer 5 ml well mixed, fresh EDTA-anticoagulated blood onto 3 ml Histopaque 1119 in a conical centrifuge tube.

   N.B. Human blood should be treated as potentially infectious and appropriate safety precautions must be taken.
2. Spin at 800 r.p.m. for 30 min.
3. Harvest leukocyte band from Histopaque-plasma interface.
4. Dilute leukocyte fraction to 10 ml using sterile PBS and centrifuge at 2500 r.p.m. for 5 min.
5. Discard supernatant and resuspend pellet in 1 ml PBS.
6. Perform cell count and adjust leukocyte concentration to 1×10^6 cells ml^{-1} using PBS.
7. Prepare duplicate cytospin slides using 100 μl of adjusted cell suspension per slide. Centrifuge at 800 r.p.m. for 3 min in a cytocentrifuge. N.B. High risk specimens (i.e. those at increased risk of containing HIV or other blood-borne viruses) should not be processed in a centrifuge. Slides should be prepared using coverslip sedimentation for 30 min in a category 3 cabinet.
8. Allow slides to air dry, then fix in 5% formaldehyde, 2% sucrose in PBS for 10 min.
9. Wash in 1% FBS in PBS for 2 min.
10. Permeabilize in 0.5% Triton X-100, 10% sucrose, 1% FBS in PBS for 5 min at room temperature.
11. Wash as in 9 above.

12. Rinse in distilled water and air dry.
13. If not stained immediately, store at 4°C overnight or at –80°C for longer periods.

B. *Staining of slides*

It is important that the slides are not allowed to dry out once staining is commenced as this may lead to staining artefacts and erroneous results.

1. Prepare a fresh 1 in 10 dilution in PBS of anti-CMV pp65 antibody (DAKO: CMV AAC10 cat # M7065).
2. Apply 40 μl of diluted antibody per slide and incubate in a humidified box at 37°C for 30 min.
3. Wash in PBS for 2 min.
4. Allow slide to drain but not dry out. Blot off excess PBS from around spot using tissue paper.
5. Apply 40 μl anti-mouse FITC antibody and incubate in a humidified box at 37°C for 30 min.
6. Wash in PBS for 2 min.
7. Rinse in distilled water, drain and mount with coverslip.
8. Examine under UV microscope using ×25 objective. Positive cells show homogeneous nuclear fluorescence, which appears very finely granular when viewed at higher magnification. Older/mature neutrophils have multilobed nuclei which may appear 'clover leaf'-like, whilst less mature neutrophils may lack the lobulated appearance. Non-specific/artefactual staining is seen in some cells and is usually cytoplasmic, either homogeneous (some mononuclear cells), or obviously granular (e.g. eosinophils).
9. Scan the whole of both slides (~200 000 PBMC) and count the number of positive cells. A positive result may be expressed as the number of positive cells per 200 000 PBMC. In renal transplant patients, 40 positive neutrophils per 200 000 PBMC is said to represent significant antigenaemia. In more heavily immunosuppressed patients (e.g. bone marrow transplant recipients) the presence of any level of antigenaemia may be significant (4).

8. Conclusions

Isolation of viruses in tissue culture remains an important technique for the diagnosis of virus infections. In theory, a single virus particle within a clinical specimen can be isolated in cell culture, and thereby expanded by many orders of magnitude, allowing accurate detection and characterization. The high degree of sensitivity and specificity inherent in virus isolation are the 'gold standard' against which new techniques should be measured. The methodology is appropriate for a wide range of both clinical specimens and different viruses. On rare occasions, its use results in the identification of unexpected, or even previously unidentified viruses: HIV and the human herpesviruses 6 and 7 are the latest in a long list of viruses first discovered in cell culture.

The two major drawbacks to this approach are that not all viruses can replicate in culture, and some viruses are rather slow to replicate. The development of new types of cell culture able to support an ever-widening array of viruses may overcome some of the former problem, whilst there are a variety of adaptations of cell culture which can speed up the process of virus identification.

References

1. Ganem, D. (1995). In *Fields virology* (ed. B. N. Fields, D. M. Knipe, and P. M. Howley). p. 2703. Raven Press, NY.
2. Shah, K. V., and Howley, P. M. (1995). In *Fields virology* (ed. B. N. Fields, D. M. Knipe, and P. M. Howley). p. 2077. Raven Press, NY.
3. Young, N. S. (1995). In *Fields virology* (ed. B. N. Fields, D. M. Knipe, and P. M. Howley). p. 2199. Raven Press, NY.
4. Tong, C. Y. W. (1997). *J. Med. Microbiol* **46**, 717.

3

Concentration and purification of viruses

P. D. MINOR

1. The need for virus concentration and purification

Virus infectivity or its neutralization by antibodies can usually be measured on non-concentrated preparations, but for many other purposes concentration and purification are essential because of the small mass of the infectious agent. For example, 1×10^6 plaque-forming units of poliovirus may have a total mass of about 1 ng, of which 250 pg will be genomic RNA. While this is sufficient for genome amplification where the sequence is known, more concentrated and purified preparations will be required for investigations such as:

- physical analysis
- genetic analysis where the genome sequence is not known
- use as an antigen or immunogen

Concentrated virus preparations will contain the virus at a higher concentration than in the starting material but still include substantial amounts of impurities such as components from host cells. These crude preparations can be used successfully in a variety of applications, including immunoprecipitation with specific antibodies, some routine antibody assays such as ELISA for known viruses, or as immunogens to produce neutralising antibodies. Possible adverse effects of non-viral impurities can include false-positive reactions with antibodies, the induction of antibodies to non-viral antigens, or the detection of non-viral nucleic acids, and some purification procedure is usual following concentration.

Purified preparations are better quality antigens and immunogens for assays or preparation of sera of high specificity (Chapter 7). The more highly purified the virus, the greater the confidence with which virus-specific components, including proteins and nucleic acid sequences, can be identified. Solution of the atomic structures of virus particles or viral proteins requires the preparation of highly purified materials in at least milligram amounts. Usually virus preparations would be concentrated before purification, but, where the virus occurs in high titre, it may be purified directly from the starting material.

1.1 Practical considerations

The first essential is a source material of adequate titre and amount to make purification possible. Examples are given in *Table 1*. On rare occasions the source may be a naturally infected specimen, as for some rotaviruses in faecal specimens where titres may be extremely high, or hepatitis C virus in human plasma where the virus cannot be grown in other systems. The supply of such natural specimens may be very limited. Plant viruses, which will not be discussed here (see Chapter 9) are usually prepared in the intact organism, while many animal viruses such as influenza, mumps, or measles virus can be grown in embryonated chicken eggs. Animal viruses can be passaged from animal to animal, although this is less common nowadays. A more suitable system is to grow the virus in some form of tissue culture, where the scale of production and the monitoring of the infectious process are easier to control (Chapter 1). The quantities of starting material required will depend on the virus and the purpose for which it is intended, but it is advisable to select a virus strain and host cell which produce a high yield wherever possible. Most polioviruses can easily be grown to infectivity titres of 10^8 pfu ml^{-1} in Hep2c cells, although the mass of virus which this represents is small, while mumps virus may be grown to 10^6 pfu ml^{-1} for some strains and cell substrates, and 10^4, i.e. a hundredfold less, for others.

Viruses vary greatly in their physical and chemical properties (1), and therefore in their lability to different treatments. The properties of the specific virus of interest may be a major factor in the details of the protocol used, although as described below many protocols are fairly generic in nature. Lipid-containing viruses, such as influenza, mumps, herpes, or hepatitis C virus, will be destroyed by organic solvents or detergents, while non-lipid-containing viruses, such as polio or SV40, may even require solvent or detergent treatment to remove contaminating cellular material in the course of purification. Viruses may differ in their sensitivity to acidic or alkaline pH:

Table 1. Example of sources and titres of material for virus purification

Virus	Source	Titre
Rotavirus	Faecal material	10^{12} particles ml^{-1}
Parvovirus B19	Blood plasma (acute phase)	10^{12} particles ml^{-1}
Hepatitis C	Plasma (acute phase)	10^6 genome equivalents ml^{-1}
Hepatitis C	Plasma (chronic phase)	10^4 genome equivalents ml^{-1}
Influenza	Allantoic fluid from chicken eggs	10^9 infectious units ml^{-1}
Mumps	Amniotic fluid from chicken eggs	10^6 infectious units ml^{-1}
Polio	Tissue culture fluid	10^8 infectious units ml^{-1}
SV40	Tissue culture fluid	10^8 infectious units ml^{-1}
Mumps	Tissue culture fluid	10^4–10^6 infectious units ml^{-1}
Rubella	Tissue culture fluid	10^4–10^6 infectious units ml^{-1}

Table 2. Properties of certain types of virus

Virus	Lipid envelope	Particle morphology	Particle size (nm)	Genome	Density ($g\ cm^{-3}$)	Sedimentation coefficient (S)
Parvovirus	no	icosahedral	20	ssDNA	1.39–1.42	110–120
Poliovirus	no	icosahedral	27	ssRNA	1.34	160
SV40	no	icosahedral	30	dsDNA	1.2[a]–1.34[b]	240
Rubella	yes	icosahedral	70	ssRNA	1.18	280
Yellow fever	yes	icosahedral	40–50	ssRNA	1.23	170–210
Hepatitis C	yes	icosahedral	40–50	ssRNA	1.09–1.11	>150
Influenza	yes	pleomorphic	~100	ssRNA	1.19	700–800
Mumps	yes	pleomorphic	~150	ssRNA	1.18–1.20	≥1000
Murine leukaemia virus	yes	spherical	80–100	ssRNA	1.16–1.18	700–800
Herpes virus	yes	complex	100–200	dsDNA	1.20–1.29	>1000

[a] in sucrose
[b] in caesium chloride

poliovirus is relatively stable at pH 4 while less so at alkaline pH, while the closely related rhinoviruses are acid-labile. The virus may tend to aggregate by virtue of the presence of cellular lipids: hepatitis A virus is routinely extracted with chloroform to break up clumps before studies of infectivity are undertaken, and hepatitis C virus in human plasma has been reported to float in solution because of its association with lipid. This has made hepatitis C difficult to purify, and it has not been consistently visualized by EM. Viruses may be large and heavy, or small and light, and may differ in density. For example, poliovirus is small, with a particle diameter of the order of 27 nm and a sedimentation coefficient of 160 S, but as it is compact and lacking a lipid layer it has a high density of 1.34 g cm^{-3}. Influenza viruses are larger, with a particle diameter of up to 120 nm and a sedimentation coefficient of 700–800 S, but have a density of only 1.19 g cm^{-3}. Virus particles may have a well defined morphology, whether lipid-containing or not, or may, like measles, mumps, and influenza, be highly pleomorphic. Examples of the properties of viruses which determine the detailed protocols to be used in purification are given in *Table 2*, including sedimentation coefficients, particle sizes, and densities (1).

The contaminants associated with a virus will depend on the preparation used, but are likely to include lipids, proteins, and nucleic acids from host cells, as well as components of the matrix such as tissue culture medium, or plasma in which the virus is suspended. The contaminants are likely to be smaller than the viruses, possible exceptions being the parvoviruses which, at 20 nm in diameter, are of comparable size to IgM molecules. Ultrafiltration can therefore be used to remove many of the smaller molecules and water, although there are disadvantages, as described below. Contaminants and viruses will be distinguishable by some physical property, such as sedimentation coefficient or density. However, the contaminants may be more or less tightly bound to the virus; this could include lipid components and cellular proteins, and while the virus can sometimes be treated to remove them, quite often this is not possible. Measles virus preparations are usually fairly impure, for example. Moreover, cellular components may be taken up within the virus particle, in which case they cannot be removed.

2. Virus concentration: general principles

The most widely applicable methods of concentrating viruses are ultracentrifugation or precipitation with either ammonium sulfate or, more usually, with polyethylene glycol and sodium chloride. Other methods are possible, including ultrafiltration, which achieves very mixed success, as the virus may tend to stick to the membrane and be lost. On an industrial scale, poliovirus is concentrated by column chromatography, which will probably only be successful in special conditions, because of adsorption of virus to the matrix. Specific methods may be devised for particular applications if necessary.

Protocol 1. Concentration of influenza virus from allantoic fluid by ultracentrifugation

Equipment and reagents

- Ultracentrifuge (e.g. Beckman L8 or equivalent)
- Sonicating water bath or probe
- Angle-head rotor to take 250 ml ultracentrifuge bottles (e.g. Beckman T19)
- Phosphate-buffered saline (PBS)

Method

1. This protocol is suitable for the concentration of large quantities of virus, for use as antigen or immunogen after further purification. One egg will yield 5–10 ml of allantoic fluid, and the yield of virus will be from 0.5 to 25 mg of purified virus per 100 eggs, depending on the growth properties of the virus strain.
2. Fill a 250 ml ultracentrifuge bottle with infected allantoic fluid.
3. Spin for 90 min at 19 000 r.p.m. (35 000 *g*) at 4 °C.
4. Decant the supernatant fluid. Refill the bottle with more allantoic fluid and repeat as required.
5. Resuspend the pellet in PBS using about 2–3 ml l^{-1} of allantoic fluid. Sodium azide (0.05% w/v) can be added as a bacteriostat if needed.
6. Transfer to any suitable thin-walled vessel, and sonicate for 1–2 min, using either a sonicating water bath or a probe. If a probe is used keep the tube in ice. Check visually for dispersal of aggregates.

Virtually all poliovirus particles can be pelleted from a tube of path length 5 cm by spinning at 100 000 g for 1 h, or from a tube of path length 12 cm by a 3 h spin at the same rate. However, only small volumes can be handled conveniently, and there may be difficulties in resuspending the virus. Precipitation with ammonium sulfate is a rapid and gentle procedure, and has been used for rhinoviruses and polioviruses. Because of the need to dilute the ammonium sulfate after pelleting, the maximum concentration achievable is only about tenfold, which is sufficient for some purposes. Recoveries are 99% or better.

Protocol 2. Concentration of poliovirus from tissue culture fluid by use of ammonium sulfate

Equipment and reagents

- Low-speed refrigerated centrifuge (e.g. Beckman J6B)
- Screw-cap centrifuge tubes
- Ammonium sulfate Anal R grade

Method

1. Add 0.4 g ammonium sulfate per ml of tissue culture fluid containing at least 1% serum to aid precipitation. Fasten cap.
2. Shake thoroughly to dissolve the crystals. The solution will become slightly acid and cloudy.
3. Spin immediately at 2000 g for 1–2 hours at 4 °C
4. Decant the supernatant and dissolve the pellet in at least 10% of the original volume of buffer.

Precipitation from tissue-culture fluid by use of polyethylene glycol and sodium chloride is widely used in some form for a variety of viruses. It can achieve a high concentration, with essentially 100% recovery, without the fluctuations in pH associated with the use of ammonium sulfate. The method can be used exactly as described (*Protocol 3*), to concentrate retroviruses from allantoic fluid from embryonated chicken eggs for purification as described later. For poliovirus, infected tissue-culture fluid from ten 800 cm^2 bottles of Hep2c cells, concentrated and purified as described later (*Protocols 6–8*), yields about 1 mg of virus.

Protocol 3. Concentration of poliovirus from tissue culture fluid by use of polyethylene glycol and sodium chloride

Equipment and reagents

- Low-speed refrigerated centrifuge (e.g. Beckman J6B)
- Screw-cap centrifuge bottles
- Polyethylene glycol 6000 (Merck)
- Sodium chloride AR
- Phosphate-buffered saline (PBS)

Protocol 3. *Continued*

Method

1. Add 2.22 g sodium chloride per 100 ml tissue culture fluid at 4 °C, and stir to dissolve.
2. When dissolved, add 7 g polyethylene glycol 6000 per 100 ml tissue-culture fluid.
3. Stir for at least 4 h, preferably overnight, at 4 °C.
4. Centrifuge at 2000 *g* for 2 h at 4 °C.
5. Decant the supernatant, and resuspend the pellet in PBS containing 2% bovine serum albumen, using 1–2 ml for each 100 ml of the initial volume of tissue culture fluid.
6. Sonicate to break up large fragments.
7. Spin at 3000 *g* for 10 min to remove solid material, retaining the concentrated virus in the supernatant.

Concentration frequently involves several steps, e.g. mumps virus concentrated by *Protocol 4* would normally be further purified on a sucrose step gradient as described later (*Protocol 10*). Typically, between 200 and 400 μl of virus at 80 HAU ml^{-1} would be produced from ten 800 cm^3 roller bottles of Vero cells and an initial volume of tissue culture fluid of one litre.

Protocol 4. Concentration of mumps virus from tissue culture fluid by use of polyethylene glycol

Equipment and reagents

- Low-speed refrigerated centrifuge (e.g. Beckman J6B)
- Screw-cap centrifuge tubes or bottles
- Polyethylene glycol 6000 (Merck)
- Sodium chloride A.R

Method

1. Dissolve 5 g of polyethylene glycol in 100 ml starting tissue-culture fluid, by stirring at 4 °C in a sealed centrifuge bottle.
2. Stir at 4 °C for 4 h.
3. Centrifuge at 1600 r.p.m. for 16 h at 4 °C.
4. Resuspend the pellet in 1 ml of phosphate-buffered saline for each 100 ml starting volume.
5. Sonicate on ice or in a sonicating water bath, using 30 s bursts until aggregates are disrupted.
6. Spin out debris at 3000 *g* for 10 min.

3. Purification of virus: identification of virus-containing fractions

As described in Section 4, the most common and successful methods of purifying virus involve ultracentrifugation on density gradients of some kind, and it is clearly essential to be able to identify the region of the gradient in which the virus is located. The following methods may be considered, but any property of the virus that can be conveniently measured can be used. The ideal assay is very rapid, and not confounded by background, either from the gradient matrix, subviral particles, or cellular contaminants.

- Visible band. Where a large amount of virus is being purified, as in the case of the material from the influenza protocol given above (*Protocol 1*), the virus is likely to be visible as a milky band in the gradient. Such bands may be visualized, if necessary, by shining a light beam from the bottom of the centrifuge tube. The method requires the use of clear centrifuge tubes, and artefactual bands of contaminants are common, although they should be fainter than the virus-containing band, and can be identified with some experience.
- Haemagglutinating activity (HA). Some viruses can bind to and cross-link red blood cells. This property can be used to identify them in gradient fractions, provided the fraction does not adversely affect the red blood cells, for example, by raising the tonicity of the assay medium to very high levels. Viruses with HA activity include influenza, certain strains of measles, certain strains of mumps, togaviruses, and even certain strains of enteroviruses. Subviral fragments may also have HA activity, but should appear well separated from the intact virus in the gradient.
- Protein assays. If there is sufficient virus, it may be possible to detect protein in the gradient fraction corresponding to the virus peaks. This may be done either by colorimetric assay or by UV monitoring, depending on the available equipment and the quantity of virus. Again, the gradient material may interfere with the assay, either non-specifically, or because of protein present in the solution which forms the gradient. Sucrose, for example, may contain low amounts of protein. The quantity of virus should be borne in mind: 10^9 infectious units of poliovirus corresponds to about 1–10 μg of protein, which is likely to be hard to detect.
- Virus-related enzyme activity. Virus-related enzymes include the polymerase associated with negative-strand viruses, and reverse transcriptase associated with retroviruses. In principle, neuraminidase activity could also be used, being associated with influenza and certain members of the Paramyxoviridae.
- Nucleic acid. If the virus is known, its nucleic acid can be sought by hybridization or PCR. The sensitivity of the assay should be adjusted so as to detect the peak on a potentially high background level of free nucleic acid.

- Infectivity. This method is applicable to all infectious agents in principle, but is likely to be very time-consuming, and must also be of the appropriate sensitivity, probably including titration of infectivity.
- Antigen detection. Where there are suitable antibodies available, such as post-infection sera, antigen-detection systems can be used, including ELISA.
- Radiolabel. One of the more convenient and widely applied methods involves the addition of radioactive amino acids, such as ^{35}S-methionine, to the medium as the virus is grown, or more usually, to reduce the use of isotope, the preparation of a separate small culture in which the virus can be labelled before addition as a marker to the bulk preparation. The labelled virus can then be detected by scintillation counting, if it is sufficiently well separated from contaminants to give a recognisable peak. Alternatively, samples can be run on a gel electrophoresis system, and the location of non-cellular proteins in peak fractions identified.

4. Purification by ultracentrifugation

4.1 Types of ultracentrifugation gradient

While non-centrifugal methods are occasionally used in purification, they are specialized in nature, and will not be described here. Essentially all centrifugation methods employ density gradients to increase the concentration of the virus in certain positions in the gradient. The solute in the gradient is usually sucrose, caesium chloride, caesium sulfate, or, less commonly, potassium sodium tartrate. These reagents are chosen for their high solubility, and the high density of the resulting solutions. *Table 3* lists the maximum densities achievable with saturated solutions of the four solutes. The distinction between weight/volume and weight/weight in the expression of concentration should be borne in mind. The appropriate gradient will be determined in part by the density of the virus of interest, as in *Table 2*, which will fall in the range 1.1–1.5 g cm^{-3}.

Table 3. Saturated solutions of solutes used in density gradient ultracentrifugation

Solute	Saturated solution w/w (g per 100 g solution)	Saturated solution w/v (g per 100 ml solution)	Density (g cm^{-3})
Sucrose	67.9	90.9	1.34
Potassium sodium tartrate	39.71	51.9	1.31
Caesium chloride	65.7	126	1.92
Caesium sulphate	64.5	129.8	2.01

4.1.1 Equilibrium density gradients

The principle underlying equilibrium density gradients is that the virus migrates until it reaches the position at which the density of the solution is the same as the density of the virus. The buoyant density may not be the same in all gradients (for example, see SV40 in *Table 2*) if one solute is able to penetrate the virion, but another solute cannot. The faster and longer the gradient is spun, the more concentrated the virus may be expected to be at the region corresponding to its buoyant density. Gradients may either be preformed, as described below, or formed during centrifugation. This may affect the required duration of the process. Lipid-containing viruses, such as retroviruses, may be purified on equilibrium sucrose density gradients, while polioviruses, being of higher density, require caesium chloride or sulfate to reach the correct density. Caesium salts are expensive, which may be a factor in deciding which procedure to follow.

4.1.2 Velocity gradients

In velocity gradients, the virus moves at a rate determined by its sedimentation coefficient and the density of the gradient matrix. The gradient is preformed, so that the leading edge of the sedimenting material is retarded compared to the trailing edge, concentrating the material in a narrower portion of the gradient.

4.1.3 Step gradients

Equilibrium and velocity gradients normally involve a continuously increasing density from the top of the gradient to the bottom. Step gradients involve layering a lower density solution onto a cushion of high density material. The sample is then layered in turn on top of the lowered density solution. Many of the impurities will then be left either at the top of the gradient, because they are of lower density or sedimentation coefficient than the virus, or sediment to the bottom of the gradient, leaving the virus at the interface between the low and high density layer. This method is especially suited to highly pleomorphic viruses such as mumps or measles, but does not produce material of a comparable purity to the other methods.

4.2 Forming gradients

A variety of procedures can be followed for forming gradients of a precision suitable for purifying viruses. Commercial apparatus is available, but an arrangement such as that shown in *Figure 1a* can give satisfactory results. A competent workshop can produce the apparatus shown in *Figure 1b*. It consists of two cylindrical chambers connected by a tap. The second chamber has an outlet tube to the centrifuge tube. The heavy component is put into the second chamber, and the light component into the first, and the connecting tap opened. As the solution flows out of the second chamber into the centrifuge

(a)

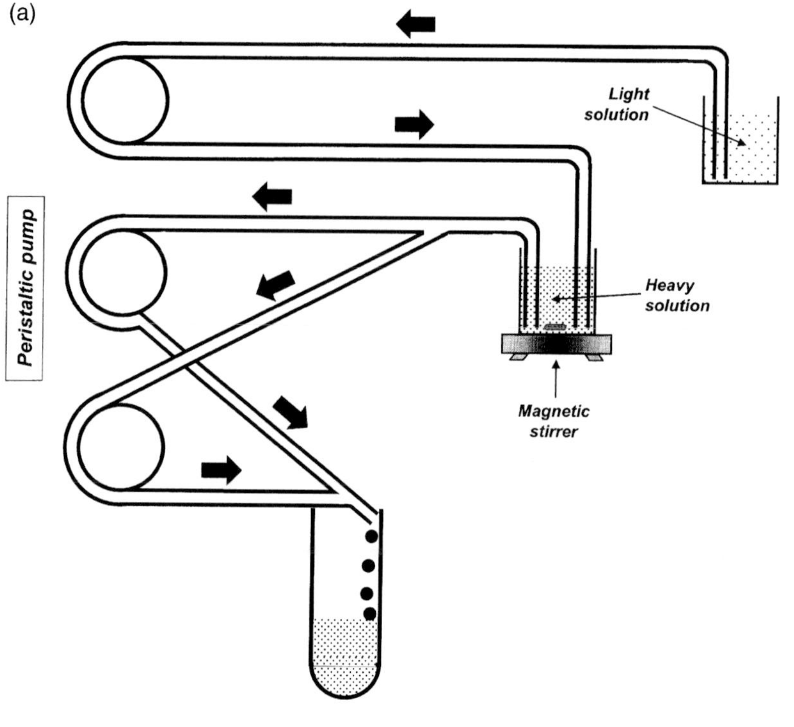

(b)

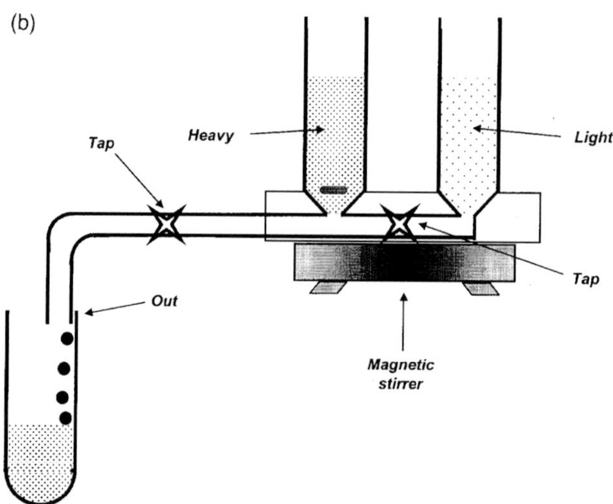

Figure 1. Gradient making apparatus. (a) use of peristaltic pump to create gradient from light and heavy solutions in separate reservoirs. (b) purpose built apparatus, use described in text.

tube it is replaced by the lighter component, the mixture being stirred the whole time, so that the contents of the mixing chamber become steadily lighter as the gradient pours.

Alternatively, it is possible to produce a gradient by layering the light component onto the heavy component in the centrifuge tube, and then, having sealed the tube top, carefully laying it on its side and standing it upright again. The gentle agitation involved in this operation is sufficient to smooth out the interface and create a gradient of a broadly acceptable nature. A second method which involves no specific apparatus involves freezing the solution in the tube, and allowing it to thaw slowly and undisturbed while the tube is positioned vertically. The higher density material will melt at a lower temperature than the rest, so that a gradient of concentration is established as the frozen melting plug floats up the tube. Gradients can be checked by measurement of the refractive index of fractions derived from them, to ensure that the procedure used is acceptable.

Protocol 5. Velocity gradient purification of influenza virus

Equipment and reagents

- Ultracentrifuge (e.g. Beckman L8)
- Swing-out rotor taking 40 ml capacity centrifuge tubes (e.g. Beckman SW28)
- 10% w/v sucrose in phosphate-buffered saline, pH 7.4
- Gradient-making apparatus
- 40% w/v sucrose in phosphate-buffered saline, pH 7.4
- Sonicating waterbath or probe

Method

1. Make 35 ml gradients, using 17.5 ml 10% sucrose and 17.5 ml 40% sucrose per gradient, in ultraclear centrifuge tubes.
2. Taking the resuspended virus concentrate from *Protocol 1*, carefully layer 2–3 ml on the top of the gradient so as to minimize mixing. This should correspond to no more than 1500 ml allantoic fluid (or the harvest from 150 to 200 eggs) per gradient.
3. Balance tubes by the dropwise addition of PBS and spin at 70 000 *g* (e.g. 22 000 r.p.m. in a Beckman ultracentrifuge using an SW28 rotor) for 60 min at 4 °C.
4. Remove the visible band with a Pasteur pipette inserted into the gradient from the top. The band should be about half-way down the tube. Pool bands from different gradients if necessary.
5. Dilute pooled bands with an equal volume of PBS, and pellet at 80 000 *g* (e.g. 25 000 r.p.m. in a Beckman SW28 rotor) for 90 min at 4 °C.
6. Decant supernatant and resuspend pellet in approximately 0.5 ml PBS per 1000 ml allantoic fluid.
7. Sonicate to remove aggregates.

Protocol 6. Velocity gradient purification of poliovirus

Equipment and reagents

- Ultracentrifuge (e.g. Beckman L8)
- Swing-out head and centrifuge tubes of 40 ml capacity (e.g. Beckman SW28)
- 10% NP40 (BDH, Poole, Dorset) in PBS
- 15% w/v sucrose in 10 mM Tris-HCl, pH 7.4, 100 mM NaCl
- 45% w/v sucrose in 10 mM Tris HCl, pH 7.4, 100 mM NaCl

Method

1. Pour 30 ml linear gradients from 15 ml each of the 15% and 45% sucrose in an ultracentrifuge tube of 35–40 ml capacity.
2. Add one tenth of a volume of 10% NP40 to the sample, whose volume should not exceed 6 ml. The material may be concentrated as described in *Protocols 2 and 3*, or may be tissue-culture fluid of high titre. The sample may be radioisotopically labelled to identify the virus peaks. The sample should clarify slightly as it is shaken with the detergent.
3. Layer the sample carefully onto the preformed gradient. Balance the tubes with liquid paraffin.
4. Centrifuge the tubes at 4°C at 80 000 *g* for 4 h (for example, at 25 000 r.p.m. in a Beckman ultracentrifuge, SW28 rotor).
5. Harvest the gradient as described (*Protocols 11–13*). Assay for the presence of virus. The infectious virus peak will be approximately two-thirds down the gradient, and will probably not form a readily visible band.

Sucrose purification of poliovirus (or other picornaviruses) by the method described in *Protocol 6* is inexpensive, and has little effect on the virus. However, some impurities are likely to remain, and if necessary the harvest from the velocity gradient can be further purified on equilibrium density gradients. As picornaviruses have a density of 1.34 g cm^{-3}, this involves the use of caesium chloride or caesium sulfate.

Protocol 7. Equilibrium density gradient purification of poliovirus on caesium chloride using preformed gradients

Equipment and reagents

- Ultracentrifuge (e.g. Beckmann L8)
- Swing-out rotor to take centrifuge tubes of 12 ml capacity (e.g. Beckmann SW41)
- Caesium chloride (Merck)
- 10 mM Tris-HCl, pH 7.4

Method

1. Prepare a solution of 40% w/w caesium chloride by dissolving 4 g of solid caesium chloride in 6 ml of 10 mM Tris-HCl, pH 7.4, and a

solution of 5% w/w caesium chloride by dissolving 0.5 g in 9.5 ml Tris-HCl, pH 7.4.

2. Pour a 10 ml 5–40% linear caesium chloride gradient in an ultracentrifuge tube of 12 ml capacity (e.g. Beckman SW41 tube), using 5 ml of each solution and a suitable gradient-making protocol. Layer the sample on the gradient in a sample volume of 1 ml. Alternatively, prepare the gradient by substituting 6 ml of sample (e.g. from a sucrose density velocity run, see above) for the 5% solution, and increasing the volume of the 40% solution to 6 ml. Clean and disinfect any gradient-making apparatus after use. Balance the tubes with liquid paraffin.
3. Centrifuge at 120 000 *g* for at least 4 h, and preferably overnight at 4 °C, in, for example, a Beckman ultracentrifuge, using an SW41 rotor at 35 000 r.p.m.
4. Harvest and assay the gradient fractions.

Protocol 8. Equilibrium density gradient purification of poliovirus on field formed caesium chloride gradients

Equipment and reagents

- Ultracentrifuge (e.g. Beckman L8)
- Swing-out rotor to take tubes of 5 ml capacity (e.g. SW Beckman SW50)
- Caesium chloride (Merck)
- NP40 (BDH Poole Dorset)

Method

1. Add solid caesium chloride to the sample to produce 0.46 g ml^{-1} of final solution volume. Thus if the tube volume is 4.5 ml, add 2.07 g of solid caesium chloride to the sample, and make up to 4.5 ml with PBS or 10 mM Tris-HCl, pH 7.5. In practice, this requires a sample volume of 3 ml or less, because of the partial specific volume of caesium chloride. Add NP40 to a final concentration of 1% v/v.
2. Centrifuge for 24 h at 4 °C at 140 000 *g* (e.g. 45 000 r.p.m. in a Beckman SW50 rotor or equivalent).
3. Harvest the gradient and assay the fractions.

The density of avian retroviruses is of the order of 1.17 g cm^{-3}, and they can therefore be purified on sucrose gradients (see *Table 3*).

Protocol 9. Equilibrium density gradient purification of avian retroviruses on preformed sucrose gradients

Equipment and reagents

- Ultracentrifuge (e.g. Beckman L8)
- Swing-out rotor to take tubes of 40 ml capacity (e.g. Beckman SW28)
- 15% w/v sucrose in 10 mM Tris, pH 7.5
- 50% w/v sucrose in 10 mM Tris, pH 7.5

Method

1. Resuspend the pellet from the polyethylene glycol concentration (*Protocol 3*) in 50–100 ml PBS.
2. Centrifuge at 4000 r.p.m for 10 min, and filter the supernatant through a 0.45 μm filter.
3. Pellet particles at 80 000 *g* for 2 h at 4 °C (for example, at 25 000 r.p.m. in a Beckman SW28 rotor). Resuspend the pellet in 1.0 ml PBS.
4. Pour a gradient using 17.5 ml 15% sucrose and 17.5 ml 50% sucrose, w/v, in 10 mM Tris-HCl, pH 7.5. Layer resuspended pellet carefully on the gradient, and balance with PBS.
5. Centrifuge at 80 000 *g* for 16 h at 4 °C (for example, 25 000 r.p.m. in a Beckman SW28 rotor).
6. Harvest, pool peak fractions, and pellet at 80 000 *g* for 2 h at 4 °C, as above.

Protocol 10. Purification of mumps virus on a sucrose step gradient

Equipment and reagents

- Ultracentrifuge (e.g. Beckman L8)
- Swing-out rotor to take 40 ml tubes (e.g. Beckman SW28)
- 25% w/v sucrose in PBS
- 60% w/v sucrose in PBS
- Sonicating water bath or probe

Method

1. Add 8 ml 60% sucrose in PBS to a 40 ml centrifuge tube (for example, a Beckman SW28 centrifuge tube).
2. Carefully layer 12 ml of 25% sucrose, in the same buffer, onto the 60% cushion.
3. Layer 10 ml of concentrated virus (*Protocol 4*) onto the formed step gradient. Balance the tubes with PBS.
4. Spin at 50 000 *g* for 2 h at 4°C (for example, 19 000 r.p.m. in a Beckman SW28 rotor).

5. Remove material at the visible interface between the low and high density sucrose layers with a Pasteur pipette from the top.
6. Dilute in PBS, and pellet in the same head at the same rate and for the same time as in step 4.
7. Resuspend the pellet in 0.2 ml PBS, and sonicate using 30 s bursts for a maximum of four cycles, to break up clumps.

Commercial apparatus is available for harvesting centrifugation gradients, but simple home-made apparatus has been found to give equally satisfactory results, and is both less expensive and easier to maintain. Three types of home-made apparatus are illustrated in *Figure 2*. The apparatus for harvesting by bottom puncture (*Figure 2a*) consists of a piece of narrow bore steel tubing pushed through a rubber bung of a size to fit the centrifuge tube concerned. The steel tubing is attached to a short piece of silicone rubber tubing.

It is easy to estimate fractions of about 0.5–2 ml by eye, and the method has been found satisfactory where peak fractions are required without any need for accurate between-gradient comparisons. An example of a gradient fractionated

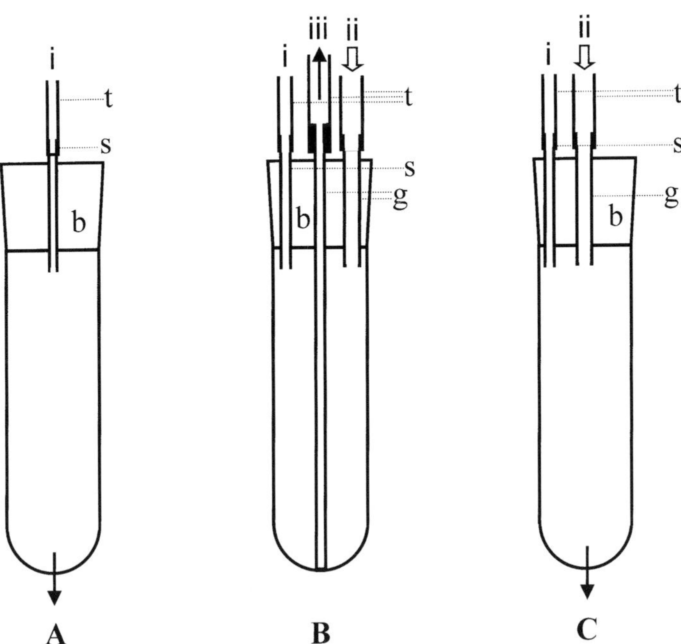

Figure 2. Simple home-made gradient harvesting apparatus. b= bung; s = stainless steel narrow bore tubing; g = glass tubing; t = silicone rubber tubing; i = bleed valve; ii = inlet for displacing fluid where used; iii = outlet tube for harvesting. For further details see text.

with this apparatus is shown in *Figure 3a*, in which radiolabelled poliovirus has been purified on a 35 ml gradient as in *Protocol 6* above, and harvested into 2 ml fractions.

Protocol 11. Harvesting by bottom puncture into estimated fractions

Equipment and reagents

- Harvesting apparatus as shown in *Figure 2a*

Method

1. Insert the bung into the centrifuge tube, held vertically in a clamp.
2. Pinch off the rubber tubing manually, and pierce the bottom of the tube with a heated needle.
3. Position a rack of tubes or other suitable collecting vessels under the hole and control the outflow by pressure on the silicone tubing.

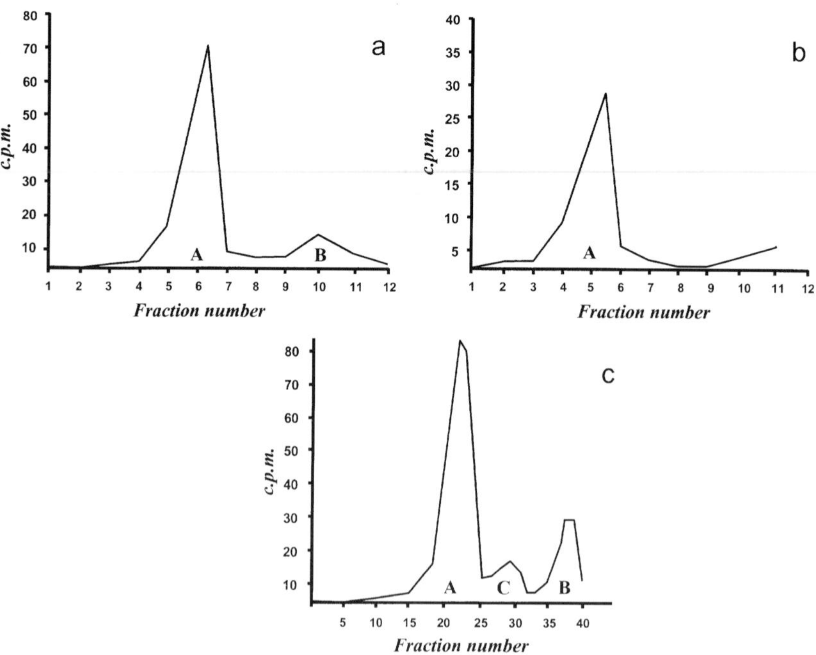

Figure 3 Examples of gradients of radiolabelled poliovirus harvested by apparatus shown in *Figure 2*. (a) [^{35}S]methionine (protein)-labelled virus harvested by the apparatus A shown in *Figure 2*. (b) [^{32}P] RNA-labelled virus harvested by apparatus B shown in *Figure 2*. (c) [^{35}S]methionine (protein)-labelled virus harvested by apparatus C shown in *Figure 2*. Peak A = infectious virus; peak B = empty capsids (RNA-free); C= uncoating intermediate. Top of gradient is to the right in all cases.

An alternative method enables gradients to be harvested without destroying the tube. While it might be thought that the procedure would severely disrupt the gradient, in practice this has not proved to be the case. An example of a gradient of ^{32}P-labelled poliovirus, harvested into 2 ml fractions by this method, is shown in *Figure 3b*. The apparatus (*Figure 2b*) consists of a rubber bung, through which are inserted a bleed tube of narrow bore stainless steel tubing attached to a short piece of silicone tubing (i); a glass inlet tube attached by silicone rubber tubing to a reservoir holding liquid paraffin (ii); and a glass tube which has been adjusted so that it nearly reaches the bottom of the tube when the bung is in place, attached to a piece of silicone rubber tubing leading to the harvest collection vessels (iii).

Protocol 12. Harvesting by displacement without tube puncture

Equipment and reagents

- Harvesting apparatus as shown in *Figure 2b*

Method

1. Clamp the centrifuge tube vertically.
2. Position clips on tubes (ii) and (iii). Carefully insert the rubber bung into the centrifuge tube.
3. Remove the clip on tube (ii), and allow the liquid paraffin to flow into the tube and overflow through the bleed tube (i).
4. Position the clip on the bleed tube (i).
5. Remove the clip on tube (iii) and collect fractions as they are displaced from the bottom of the tube.

Virus may also be harvested from gradients into known volumes by bottom puncture. The apparatus (*Figure 2c*) consists of a rubber bung as above, a bleed tube of steel attached to a short section of silicone rubber tubing (i), and a glass inlet tube attached to an adjustable repeating syringe able to deliver 0.5–2 ml of liquid paraffin. Virus fractionated in this way into 0.5ml fractions from a 36 ml SW28 gradient of [^{35}S]methionine-labelled poliovirus is shown in *Figure 3c*. Three peaks are distinguishable: A is the infectious virus peak (160 S), B is the empty capsid peak (80 S) and C is an uncoating intermediate (130 S).

Protocol 13. Harvesting into known volumes by bottom puncture

Equipment and reagents

- Harvesting apparatus as shown in *Figure 2c*

Method

1. Clamp the centrifuge tube vertically.

Protocol 13. *Continued*

2. Insert the rubber bung.
3. Pump liquid paraffin into the tube until it overflows through the bleed tube. Clamp the bleed tube.
4. Pierce the bottom of the centrifuge tube with a heated needle.
5. Displace the gradient into collecting vessels with liquid paraffin delivered by the syringe into known volumes.

5. Assessment of purity

The purity of the final preparation may be assessed qualitatively by a variety of methods, depending on the virus, the amount present, and the impurities anticipated. One method is the use of negative staining for electron microscopy. An example of a purified poliovirus preparation is shown in *Figure 4*. This requires the availability of the appropriate equipment, which may not always be available or convenient. An alternative is the use of polyacrylamide

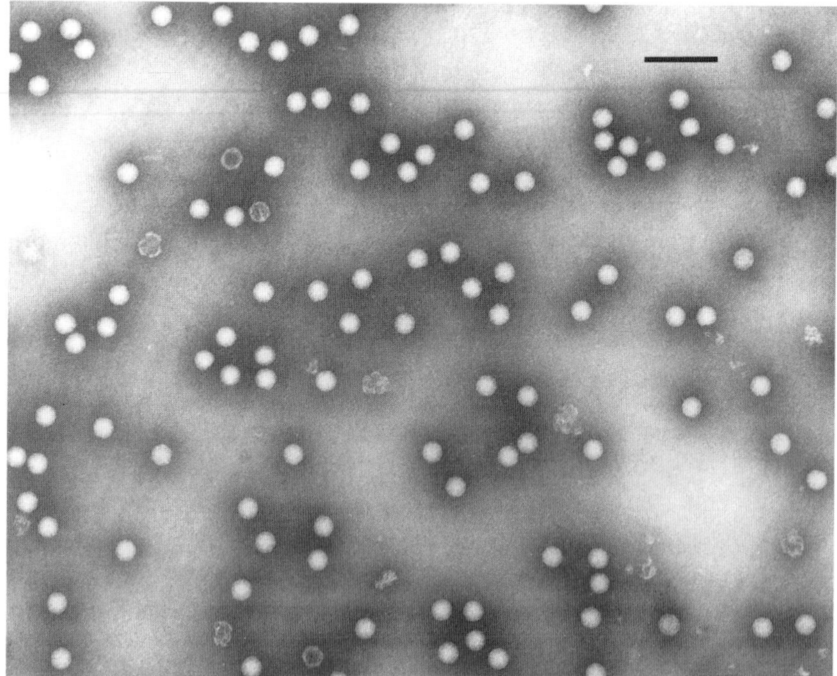

Figure 4. Negatively stained electron micrograph preparation of purified infectious virus. Scale bar = 100 nm.

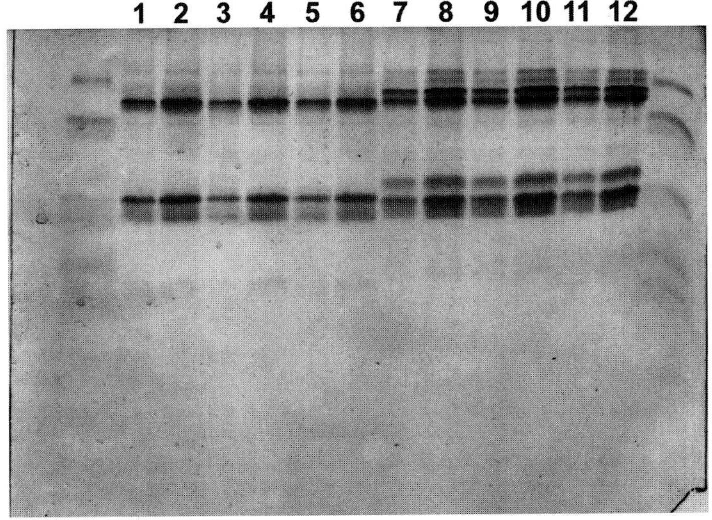

Figure 5. Coomassie blue-stained 10% polyacrylamide gel of purified preparations of influenza virus. Lanes 1–6 A/Beijing/262/95 H1N1 virus. Lanes 7–12. B/Beijing/243/97 virus.

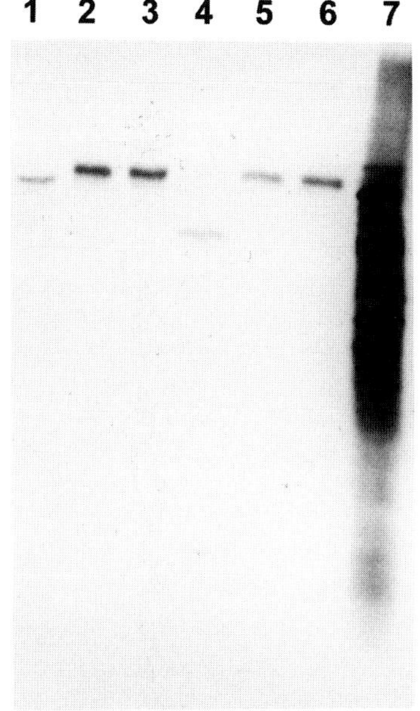

Figure 6. Immunoprecipitation with mumps-specific monoclonal antibodies from [^{35}S]methionine-labelled mumps virus concentrated and purified as described in the text. Lane 1: anti nucleoprotein antibody; Lanes 2, 3, 5, 6: different anti-haemagglutinin-neuraminidase antibodies; Lane 4: anti-fusion protein antibody.; Lane 7 shows the preparation of mumps virus after concentration and purification.

gel electrophoresis, depending on the amount of material available. An example of a stained 10% polyacrylamide gel of a number of influenza preparations produced by the protocols described here is shown in *Figure 5*. Variations on this theme might include Western blotting of gels using antisera specific for the virus or for the uninfected cell type in which it was grown. In many cases the purity of the preparation may be poor, although it may still be readily usable. An example of a [^{35}S]methionine-labelled purified mumps preparation is shown in *Figure 6*, with the specific bands precipitated from it used to identify the specificity of several monoclonal antibodies.

References

1. Murphy, F. A., Fauquet, C. M., Bishop, D. H. L., Ghabrial, S. A., Jarvis, A. W., Martelli, G. P., Masp, M. A., and Summers, M. D. (1995). *Archives of Virology*, Supplement **10**.

4

Assays for virus infection

J. GRAY

1. Introduction

The laboratory diagnosis of virus infections utilizes a wide range of techniques, including virus isolation (Chapter 2), the detection of virus antigens and specific antiviral antibodies and the amplification and detection of virus genomes. This chapter will outline the principal methods used for the diagnosis of virus infections, and describe in detail those which are used most commonly.

2. Identification of viruses isolated in cell culture

2.1 Cytopathic effects

This section will outline methods used to identify and quantify viruses grown in cell culture, since detailed methods for the isolation of viruses are described in Chapter 2. Many of the cytopathic effects (CPE) seen in virus-infected cell cultures are characteristic of specific viruses or virus groups (*Table 1*), and the ability to recognize a specific CPE is the important first step in virus identification. Not all viruses growing in cell culture produce a CPE, and other methods have been devised for their identification (Section 2.7).

2.2 Virus infectivity assays

The object of these assays is to measure the number of infectious particles in a suspension of viruses. This is done either by plaquing viruses on monolayer cell cultures in order to reveal and enumerate plaques which originate from a single infectious virus particle (pfu) (*Protocol 1*), or by determining the highest dilution of the virus suspension which produces a CPE in 50% of the cell cultures inoculated ($TCID_{50}$) (*Protocol 2*). The $TCID_{50}$ value is calculated by use of the Reed-Muench formula or the Karber method (1). Calculation of the $TCID_{50}$ is more commonly used to determine the dose of virus required in a neutralization test, although plaque assays give a more precise measurement of virus infectivity.

Table 1. Cytopathic effects caused by viruses growing in cell cultures

Virus	Susceptible cell line	Appearance of cytopathic effect
Herpes simplex virus	MRC-5, human amnion, Vero	24-48 h after inoculation grape-like clusters of round highly refractile cells can be seen
Cytomegalovirus	MRC-5	Focal cytopathic effect involving the formation of giant cells and generally restricted to the interior of the cell sheet, usually appearing at least 6 days after inoculation
Varicella zoster virus	MRC-5	Similar CPE to that seen with CMV in the early stages but tends to be found near the margin of the cell sheet
Adenovirus	Human amnion, rhesus monkey kidney, HEp2, MRC-5	CPE first seen at the margin of the cell sheet with characteristic round refractile clusters of cells sometimes joined by cytoplasmic bridges
Respiratory syncytial virus	HEp2, rhesus monkey kidney, MRC-5	Characteristic CPE with the formation of syncytia, usually dark and raised, 2–7 days after inoculation
Measles virus	Vero	CPE develops between 48 h and 15 days and consists of large syncytia, very flat in appearance
Reovirus	Rhesus monkey kidney	Cell sheet takes on a ragged appearance with strands of cells detaching from the cell culture tube
Enterovirus	Rhesus monkey kidney, human amnion, MRC-5, Hep2	Cells become rounded, refractile and ultimately shrink before detaching from the cell surface
Rhinovirus	MRC-5	Sometimes difficult to distinguish from enterovirus CPE but rounded cells tend to be smaller, the cell sheet appears "cleaner" and the CPE develops more slowly
Influenza viruses A and B	Rhesus monkey kidney, MDCK	CPE may be mistaken for enterovirus or adenovirus CPE. It is generally scruffy in appearance
Parainfluenza virus type 2	Rhesus monkey kidney	Typical well demarcated syncytia, often round, dark and granular
Parainfluenza virus type 3	Rhesus monkey kidney	Rounding at the edge of the cell sheet with long spike-like processes
Parainfluenza virus type 4	Rhesus monkey kidney	As with parainfluenza virus type 2 and mumps virus but usually slower to develop
Mumps virus	Rhesus monkey kidney	CPE similar to that seen with parainfluenza virus type 2

Protocol 1. Determining the infectivity of enteroviruses, using a plaque assay

Equipment and reagents

- Maintenance medium: Balanced salt solution 199 (Gibco) containing 2% fetal bovine serum (FBS) and antibiotics (2)
- Virus suspension
- Confluent monolayers of Vero cells in six-well tissue culture plates (Nunc)
- CO_2 incubator at 37 °C
- 1% (w/v) agarose
- 2× nutrient overlay solution: double strength BSS 199 (Gibco) containing 4% FBS
- Methanol
- 0.5% Crystal violet in methanol-water (20:80, v/v)

Method

1. Prepare tenfold dilutions of the virus suspension in maintenance medium, changing pipettes between dilutions to avoid carry-over.
2. Remove the growth medium from the confluent monolayers of Vero cells.
3. Add 0.1 ml of each dilution of virus suspension to each of two wells, and incubate the plates at 37 °C in an atmosphere of 5% CO_2 for 1–2 h to allow virus absorption.
4. Bring 2× nutrient overlay solution to 46 °C and add an equal volume of 1% agarose at 46 °C.
5. Add 3 ml of the overlay solution containing agarose to each well, allow it to set, and then incubate for 5 days in a CO_2 incubator at 37 °C.
6. Remove the overlay and fix the cells in methanol for 10 min.
7. Pour off the methanol and stain the cell sheets with 0.5% crystal violet in methanol-water (20:80, v/v) for 20 min.
8. Pour off the stain. Rinse the cells in tap water and leave to dry.
9. Count the number of plaques at each dilution and calculate the pfu per 0.1 ml.

Protocol 2. Determining the $TCID_{50}$ of an enterovirus suspension

Equipment and reagents

- Maintenance medium (*Protocol 1*)
- Virus suspension
- CO_2 incubator
- Confluent monolayers of Vero cells in cell roller culture tubes (Sterilin)

Method

1. Prepare tenfold dilutions of the virus suspension in maintenance medium.

Protocol 2. *Continued*

2. Remove the growth medium from the confluent monolayers of Vero cells in cell culture tubes, and replace with 0.9 ml of maintenance medium.
3. Add 0.1 ml of the virus dilution to each of four Vero cell roller culture tubes, and incubate at 37 °C for 5 days. Include four uninoculated cell cultures as controls.
4. Examine the tubes for characteristic enterovirus CPE, and record the number of infected and uninfected cell cultures at each virus dilution.
5. Determine the virus infectivity using the Karber method (*Table 2*).

Table 2. Virus infectivity determination: Karber method
The method is best explained using a case study. To describe the procedure, the following data will be used:

Virus dilution	CPE present Tube				Proportion of tubes infected
	1	2	3	4	
10^{-1}	+	+	+	+	4/4 = 1
10^{-2}	+	+	+	+	4/4 = 1
10^{-3}	+	+	+	–	3/4 = 0.75
10^{-4}	+	+	–	–	2/4 = 0.5
10^{-5}	+	+	–	–	2/4 = 0.5
10^{-6}	–	–	–	–	0/4 = 0.0
					sum of proportions = 3.75

Calculate the $TCID_{50}$ using the following formula:

$$\text{Log } TCID_{50} = L - d\,(s - 0.5)$$

where L is the log of the lowest dilution (= -1 in this case), d is the difference between dilution steps (= 1 in this case), s is the sum of the proportion of positive tubes (= 3.75 in this case).

For the sample data given above:

$$\text{Log } TCID_{50} = -1 - 1\,(3.75 - 0.5)$$
$$= -4.25$$
$$TCID_{50} = 10^{-4.25}$$
$$100\ TCID_{50} = 10^{-2.25}$$

Therefore the virus suspension must be diluted between 1 in 100 (10^{-2}) and 1 in 1000 (10^{-3}). The exact dilution is obtained from the antilog of 0.25 (from $10^{-2.25}$).

antilog 0.25 = 178

Therefore the dilution of virus suspension that gives 100 $TCID_{50}$ is 1 in 178.

2.2.1 Neutralization tests

Any effects of virus growth can be neutralized with virus-specific antibodies. These include:

- growth leading to pathological changes in a susceptible animal host
- growth in cell culture causing a CPE in susceptible cells (*Table 3*)

- haemadsorption and/or haemagglutination of red blood cells through their interaction with virus haemagglutinins

Table 3. Poliovirus neutralisation tests

Poliovirus antiserum	Specimen 1 CPE present	Specimen 2 CPE present
poliovirus types 1, 2 and 3	no	no
poliovirus type 1	yes	yes
poliovirus type 2	no	yes
poliovirus type 3	yes	yes
poliovirus types 1 and 2	no	no
poliovirus types 1 and 3	yes	yes
poliovirus types 2 and 3	no	yes
no antiserum - virus control	yes	yes
no antiserum, no virus - cell control	no	no
Result	poliovirus type 2	poliovirus types 1 and 2

Neutralization tests in cell culture are performed using virus-specific hyperimmune sera (Chapter 2). The unknown virus, at a concentration of 100 tissue-culture infectious doses ($TCID_{50}$) or 100 plaque forming units per 0.1 ml (pfu), is incubated for 1 h at room temperature with an equal volume of specific antisera, at a dilution previously determined to neutralize 100 $TCID_{50}$ or 100 pfu per 0.1 ml of virus. The mixture is inoculated into susceptible cell cultures after incubation. Appropriate virus and cell controls are included, and the cell cultures are observed daily for the appearance of CPE, plaques, or haemadsorption after the addition of red blood cells. The appropriate virus concentration for use in the neutralization test is determined either in a plaque assay (*Protocol 1*), or a virus infectivity titration (*Protocol 2*).

2.3 Immunofluorescence

Viruses growing in cell culture can be identified by direct or indirect immunofluorescence with virus-specific antibodies (Chapter 2). In direct immunofluorescence assays, type-specific antibodies labelled with fluorescein isothiocyanate (FITC) are reacted with virus-infected cells fixed to microscope slides. Specific antigen-antibody interactions are detected by means of an epifluorescence microscope. Light in the 400–500 nm range (blue light) excites the typical green fluorescence of fluorescein. In indirect immunofluorescence, the specific antibody is unlabelled. Binding of antibody to antigen is detected after a second incubation with a species-specific antibody labelled with FITC (*Figure 1*).

2.4 Electron microscopy

The characteristic morphology of viruses allows their identification by electron microscopy (EM), and specific serotypes can be identified by using

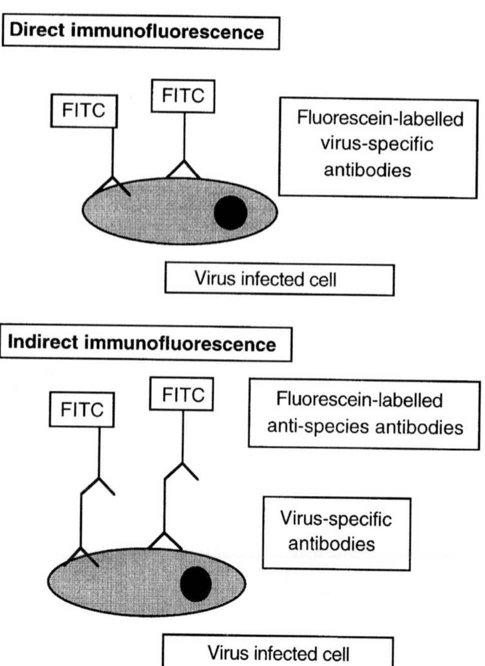

Figure 1. Direct immunofluorescence, in which virus antigens are detected with a fluorescein-labelled virus-specific antibody. Indirect immunofluorescence, in which a second fluorescein-labelled species-specific antibody is used to detect the virus-antibody interaction.

type-specific antibodies in immune EM. Cell culture-grown virus complexed with specific antibody is captured onto protein A-coated EM grids. Immune complexes comprising virus particles clumped by specific antibodies are indicative of the virus type. For more details, see Chapter 5.

2.5 Other methods of identifying viruses growing in cell culture

Haemadsorbing viruses, including influenza viruses A and B, parainfluenza viruses 1, 2, 3, 4a, and 4b, mumps virus, and the monkey virus SV5 can be detected in baboon or rhesus monkey kidney cell cultures by their ability to haemadsorb guinea pig or human group O red blood cells (*Table 4*). Influenza viruses A and B and measles virus can be detected in cell-culture fluid by their ability to haemagglutinate fowl or monkey red blood cells respectively (Chapter 2).

The complement-fixation test (CFT) can be used to identify viruses grown in cell culture. Virus antigen bound to specific antibodies fixes complement, thus preventing the lysis of sheep red blood cells coated with anti-sheep red blood cell antibodies (3).

Table 4. Haemadsorption of myxo- and paramyxoviruses with human group O red blood cells[a]

Virus	Haemadsorption at 4°C	Haemadsorption at room temperature
Influenza A	+++	++
Influenza B	+++	++
Parainfluenza 1	+++	−
Parainfluenza 2	+++	+++
Parainfluenza 3	+++	−
Parainfluenza 4a	+/−	+++
Parainfluenza 4b	+/−	+++
Mumps	+++	+++
Simian virus 5	+++	−

[a]In influenza virus infected cell cultures where there is a high titre of haemagglutinin in the cell culture fluid the red blood cells may be aggregated to such an extent that there is little or no haemadsorption.

2.6 Detection of virus pre-cytopathic effects

Virus-specific antigens in detectable quantities are often present in infected cells prior to the appearance of CPE. This property has been used to develop rapid diagnostic tests for detecting viruses growing in cell culture. Cytomegalovirus (CMV) may take a minimum of 6 days to produce a cytopathic effect, but virus-specific antigens (CMV immediate early and early antigens) can be detected 20–40 h after inoculation (4) (Chapter 2).

2.7 Detection of viruses not producing a cytopathic effect

Viruses may grow in cell culture without producing a CPE. Viruses that replicate slowly, are non-lytic, or are present in very low concentrations, often fail to produce a characteristic CPE. Cytomegalovirus at low concentrations may take up to 6 weeks to produce a CPE, and rubella virus will replicate in rabbit kidney cells but only produce a CPE when passaged to Vero cells after 28 days. Many of the techniques mentioned above, such as immunofluorescence, electron microscopy, and CFT, can be used to detect viruses growing in the absence of CPE. These techniques are labour-intensive, and expensive if applied to all CPE-negative cultures.

2.7.1 Interference

A more efficient way of detecting viruses in the absence of a CPE is to use interference. A virus infecting a cell will make that cell non-permissive to a second virus, due either to the induction of interferon synthesis, or sequestering of host cell receptors. Therefore, after a suitable period of incubation, cell cultures are challenged with a fast-growing, CPE-producing, unrelated virus. For rubella virus detection an enterovirus is often used. Cell sheets with an enterovirus CPE can be discarded as negative for rubella virus. Viruses

Table 5. Methods used to detect virus antigens in clinical specimens

Virus	**Clinical sample**	**Method used**
Influenza virus	NPA	Direct IF
Parainfluenza virus	NPA	Direct IF
Respiratory syncytial virus	NPA	Direct IF
Adenovirus (respiratory)	NPA	Direct IF
Adenovirus (gastrointestinal)	Faeces	Antigen-capture ELISA, particle agglutination test
Rotavirus	Faeces	Antigen-capture ELISA, particle agglutination test
Astrovirus	Faeces	Antigen-capture ELISA
Herpes simplex virus	Lesion swab/scrape	Direct IF, antigen-capture ELISA
Varicella zoster virus	Lesion swab/scrape	Direct IF
Cytomegalovirus	Buffy coat	Direct IF

detected through an absence of enterovirus CPE can be identified by immunofluorescence, or passaged into fresh cultures after the enterovirus has been neutralized with specific antisera.

3. Virus antigen detection

Virus antigens can be detected directly from clinical specimens, using a variety of immunological techniques. Direct immunofluorescence, particle agglutination, and antigen capture ELISAs utilizing virus-specific monoclonal antibodies have been developed to detect a wide range of viruses (*Table 5*).

3.1 Immunofluorescence

A nasopharyngeal aspirate (NPA) is the specimen of choice for detecting respiratory virus antigens in patients with virus infection of the upper or lower respiratory tract. The specimen contains an abundance of potentially infected ciliated columnar epithelial cells.

Protocol 3. Detection of respiratory virus antigens, using direct immunofluorescence

Equipment and reagents

- Hanks lactalbumin (Gibco)
- Bovine serum albumin: fraction V (Sigma)
- Bench top centrifuge
- Conical centrifuge tubes (Costar)
- PBS (Oxoid)
- 12-well PTFE microscope slides (ICN Biomedicals)
- Acetone (BDH)
- FITC-labelled antibodies to influenza viruses A and B, parainfluenza viruses 1, 2, and 3, and respiratory syncytial virus containing Evans blue (Dako)
- Tween 20 (Sigma)
- Coverslips (BDH)
- Mounting medium (Sigma)
- Epifluorescence microscope with a 50x objective (Leitz)

Method

1. Mix 0.5 ml of the NPA with 2 ml of Hanks lactalbumin, containing 1% bovine serum albumin, in a conical centrifuge tube.
2. Centrifuge at 2000 r.p.m. for 10 min, and remove the supernatant. The supernatant is used to inoculate appropriate cell cultures.
3. Add 5 ml PBS (pH 7.2) to the cell deposit, resuspend the cells, and centrifuge at 2000 r.p.m. for 10 min.
4. Repeat the wash step if the specimen contains mucus.
5. Resuspend the cell deposit in 0.5 ml of PBS, and add one drop with a Pasteur pipette to each of six wells on a 12-well PTFE-coated microscope slide, ensuring the drop is spread over the entire well.
6. Dry the slide in air, and fix the cells in acetone for 15 min at room temperature.
7. Add 25 μl of the appropriate FITC-labelled antibody conjugate[a] to a well on the slide, including influenza viruses A and B, parainfluenza viruses 1, 2, and 3, and respiratory syncytial virus.
8. Incubate the slide at 37°C for 30 min.
9. Rinse the slide with PBS containing 0.025% Tween 20 for 15–30 s.
10. Dry in air and mount with a coverslip.
11. Examine each well for the characteristic apple-green fluorescence, using an epifluorescence microscope.
12. Interpretation:

Influenza A and B: Fluorescence is nuclear, cytoplasmic, or both. Nuclear staining is uniformly bright, and cytoplasmic staining is punctate with large inclusions.

Parainfluenza types 1, 2, and 3: Fluorescence is confined to the cytoplasm, and is punctate with irregular inclusions.

Respiratory syncytial virus: Fluorescence is seen in the cytoplasm, and is punctate with small inclusions.

Negative cells: Exhibit a dull red staining in the cytoplasm, and maroon staining in the nucleus.

[a] The FITC-labelled antibody conjugate should contain Evans blue as a counterstain.

3.2 Particle agglutination

Antibodies coupled to red blood cells, latex particles, or gelatin particles can be used to detect virus antigens (5). A smooth suspension of particles becomes particulate with visible agglutination in the presence of a specific antigen-antibody reaction. Control particles coated with non-viral antibodies derived from the same animal species should be used to detect non-specific agglutination.

3.3 Antigen-capture ELISA

Virus-specific antibodies immobilized on the plastic surface of microtitre wells can be used to capture specific virus antigens. The captured virus is then detected with a second specific antibody (produced in a different animal species or directed against a different epitope) labelled with an enzyme whose presence initiates a colour change in an enzyme substrate-chromogen mixture. An assay is given here for a rotavirus antigen-capture ELISA, which is particularly useful, since these common viruses are frequently difficult to isolate from clinical samples (Chapter 2).

Protocol 4. Rotavirus antigen-capture ELISA

Equipment and reagents

- Flat-bottomed microtitre plates (Becton Dickinson)
- Rabbit anti-group A rotavirus antibodies (Dako)
- 0.1 M sodium carbonate-bicarbonate buffer, pH 9.6
- 0.1 M PBS, pH 7.2 (Oxoid)
- Tween 20 (Sigma)
- Microtitre plate washer (Denley)
- Skimmed milk powder (Oxoid)
- BSS 199 (Sigma)
- Faecal samples
- Incubator at 37 °C
- Mouse anti-rotavirus monoclonal antibody (Amrad)
- HRPO-labelled anti-mouse antibodies (Dako)
- Tetramethylbenzidine (Sigma)
- Hydrogen peroxide (Sigma)
- Substrate buffer (pH 6.0) (8.2 g sodium acetate, 19.21 g citric acid, made up to 1 l with distilled water)
- 2 M sulfuric acid
- Spectrophotometer with 450 nm wavelength filter.

Method

1. Coat each well of a flat-bottomed microtitre plate with 100 μl of a predetermined concentration of rabbit-anti-Group A rotavirus antibodies in carbonate-bicarbonate buffer overnight at 4 °C.
2. Wash the plate three times with PBS containing 0.05% Tween 20. In each wash, completely fill the empty wells.
3. Add 200 μl of PBS containing 5% skimmed milk powder, and incubate at 37 °C for 30 min to block the uncoated sites.
4. Remove the blocking agent by aspiration.
5. Add 100 μl of a 10% extract of faeces in BSS 199 (Sigma) to the well, and incubate at 37 °C for 2 h.
6. Wash the plate five times with PBS-Tween as before.
7. Add 100 μl of mouse anti-rotavirus antibody to each well, and incubate at 37 °C for 2 h.

8. Wash the plate five times with PBS-Tween as before.
9. Add 100 μl of HRPO-labelled anti-mouse antibodies to each well, and incubate at 37 °C for 2 h.
10. Wash the plate five times with PBS-Tween as before.
11. Add 100 μl tetramethylbenzidine (0.1 mg ml^{-1}) containing 0.014% hydrogen peroxide in substrate buffer (pH 6.0) to each well.
12. Incubate the plate for 30 min at room temperature.
13. Add 100 μl of 2 M sulfuric acid to each well to stop the reaction, and measure the optical density values at a wavelength of 450 nm.

4. Virus genome detection

Genome detection offers the possibility to detect, with a high degree of sensitivity and specificity, virus nucleic acid sequences directly from clinical specimens, or to identify viruses grown in cell culture. Techniques have been developed to extract virus DNA and RNA from clinical samples, to reverse transcribe virus RNA to provide a DNA copy (cDNA), and to amplify DNA by means of DNA synthesis via oligonucleotide primer extension, as in the polymerase chain reaction (PCR) (6), and nucleic acid sequence-based amplification (NASBA) (7), or to amplify a nucleic acid hybridization signal as in the branched DNA (bDNA) assay (8).

4.1 Nucleic acid extraction

There are many different protocols for the extraction of DNA and RNA from clinical samples or infected cell cultures, and the choice will depend on the complexity of the sample, their efficiency in removing inhibitors of PCR, and their ease of use. Treatment with the enzyme proteinase K followed by phenol-chloroform extraction (9), or with chaotropic agents such as guanidinium isothiocyanate followed by immobilization of nucleic acids on a silica solid-phase matrix (10), are both methods widely used for the extraction of virus DNA and RNA from clinical samples.

Protocol 5. Preparation of reagents for silica-guanidinium isothiocyanate nucleic acid extraction

Equipment and reagents

- Guanidinium isothiocyanate (Fluka)
- 0.1 M Tris-HCl, pH 6.4
- 0.2 M EDTA, pH 8.0
- Triton X-100 (Sigma)
- Silicon dioxide (Sigma)
- Concentrated HCl
- Diethylpyrocarbonate (DEPC) (Sigma)
- Fume cupboard

Protocol 5. *Continued*

Method

1. To make buffer L6, add 120 g of guanidinium isothiocyanate to 100 ml of 0.1 M Tris-HCl, pH 6.4, and 22 ml of 0.2 M EDTA, pH 8.0, and 2.6 g of Triton X-100; stir overnight in the dark to dissolve. Store in the dark.
2. To make buffer L2, add 120 g of guanidinium isothiocyanate to 100 ml of 0.1 M Tris-HCl, pH 6.4, and stir overnight in the dark to dissolve. Store in the dark.
3. To make size-fractionated silica, add 60 g of silicon dioxide to 500 ml of distilled water in a measuring cylinder, and allow to stand for 24 h at room temperature. Discard 430 ml of supernatant, and resuspend solids in 500 ml of distilled water. Allow to stand at room temperature for 5 h, and discard 440 ml of supernatant. Add 600 μl of concentrated HCl, pH 2.0, and aliquot into 1.5 ml volumes. Sterilize by autoclaving and store in the dark.
4. To make RNase-free distilled water, add 100 μl of diethylpyrocarbonate (DEPC) (Sigma) in a fume cupboard to 100 ml of tissue culture-grade distilled water, and incubate for $>$12 h at 37 °C. Sterilize by autoclaving.

Protocol 6. Extraction of RNA from faeces using silica-guanidinium isothiocyanate

Equipment and reagents

- Screw-capped microcentrifuge tubes (Sarstedt)
- Microcentrifuge (Sigma)
- Vortex mixer (Sigma)
- L6 buffer (*Protocol 5*)
- L2 buffer (*Protocol 5*)
- Size-fractionated silica (*Protocol 5*)
- Ethanol (BDH)
- Acetone (BDH)
- Dry heating block (Grant)
- RNase-free distilled water (*Protocol 5*)
- RNasin (Promega)

Method

1. Add 100 μl of a 10% faecal extract to 500 μl of L6 buffer and 10 μl of size-fractionated silica in a screw-capped 1.5 ml microcentrifuge tube.
2. Vortex the tube for 10 s and incubate, shaking at room temperature, for 15 min.
3. Centrifuge the tube for 15 s at 13 000 r.p.m., and collect the supernatant for disposal[a].
4. Wash the pellet twice with 500 μl L2 buffer, twice with 500 μl of 70% ethanol, and once with 500 μl of acetone. Centrifuge for 15 s at 13 000 r.p.m. after each wash, and collect the supernatant for disposal.

5. Remove the acetone, place the tube, with the lid open, at 56 °C in a dry heating block for 5 min.
6. Elute the nucleic acid from the silica by adding 30 μl of RNase-free distilled water containing 20–40 units of RNasin (Promega), vortex the tube, and incubate at 56 °C for 15 min.
7. Centrifuge the tube at 13 000 r.p.m. for 2 min, and extract the supernatant[b]. The extracted RNA can be stored at 4 °C overnight, or at −70 °C for longer.

[a] Guanidinium isothiocyanate under acidic condition will release cyanide gas. Discard any waste into an equal volume of 10 N NaOH.
[b] Care must be taken not to contaminate the extract with silica.

4.2 Genome detection assays

4.2.1 Polyacrylamide gel electrophoresis

Polyacrylamide gel electrophoresis (PAGE) is used to characterize viruses with segmented RNA genomes, such as rotavirus, into electropherotypes, allowing the detection of genome rearrangements (11). Also, the analysis of restriction endonuclease fragments of DNA virus genomes by PAGE allows one to explore the relatedness of clusters of infections with herpes simplex virus (12), cytomegalovirus (13), or adenovirus (14), which are suspected of having a common source.

Protocol 7. Polyacrylamide gel electrophoresis of rotavirus RNA

Equipment and reagents

- Acrylamide (Sigma)
- bis-acrylamide (Sigma)
- 1.5 M Tris, pH 8.8
- Sodium dodecyl sulfate (SDS) (Sigma)
- N,N,N′,N′-tetramethylethylenediamine (TEMED) (Sigma)
- 20% ammonium persulfate (APS) (Sigma)
- Two 20 cm^2 glass plates
- Ethanol (BDH)
- Ether (BDH)
- Water-saturated *sec*-butanol (BDH)
- Electrophoresis comb and spacers (Anachem)
- Vertical electrophoresis tank (Anachem)
- 10× running buffer (30 g Tris and 144 g glycine to 1 l of distilled water)
- Loading buffer (2.5 ml 0.5 M Tris, pH 6.8, 4.0 ml glycerol, 1.0 ml SDS, 2.5 ml water, and 0.1 ml of 0.1% bromophenol blue).
- Fixing solution (125 ml methanol, 50 ml glacial acetic acid, and 325 ml distilled water).
- 0.19% silver nitrate
- Developer (14 g NaOH and 8 ml 37% formaldehyde in 1 l distilled water)

Method

1. Prepare a resolving gel by mixing 12.5 ml acrylamide (AA)-bisacrylamide (bisAA) (30% AA, 0.8% bisAA), 10 ml 1.5 M Tris, pH 8.8, 17.5 ml distilled water, and 400 μl of 10% SDS.
2. Degas for 20–30 min.

Protocol 7. *Continued*

3. Add 50 μl of TEMED and 75μl of 20% APS.
4. Fill between two 20 cm^2 glass plates cleaned with 70% ethanol and ether, and containing 1 mm thick spacers at the bottom and left and right hand sides.
5. Overlay with 2 ml of H_2O-saturated *sec*-butanol, and leave for at least 30 min to allow polymerization.
6. Prepare the collecting gel by mixing 1.5 ml acrylamide-bisacrylamide, 2.5 ml 0.5 M Tris, pH 6.8, 6.0 ml distilled water, and 100 μl of 10% SDS.
7. Degas for 20–30 min.
8. Pour off butanol from the resolving gel, and wash with distilled water three times.
9. Add 30 μl TEMED and 30 μl of 20% APS to the collecting gel, and pour 6 ml between the glass plates.
10. Immediately position the comb and leave for at least 30 min.
11. Remove the comb and bottom spacer.
12. Add the running buffer, containing 1 ml of 10% SDS per 100 ml buffer, to the bottom reservoir of the electrophoresis tank.
13. Inset the glass plate into the tank, and add running buffer to the top reservoir.
14. Remove air bubbles from the bottom of the gel, and wash the slots with running buffer.
15. Load the extracted RNA, reconstituted in 15 μl of loading buffer.
16. Run the gel at 1 mA cm^{-1} for 18–20 h.
17. Discard the running buffer, and remove the gel from the glass plates.
18. Fix the gel for 30 min in fixing solution.
19. Remove the fixative and stain the gel with 0.19% silver nitrate solution for at least 30 min at room temperature in the dark.
20. Wash the gel three times, 15 s each wash, with distilled water.
21. Add the developer, and stir for 1 min. Leave for up to 10 min, checking for visualization of RNA segments.
22. Decant the developer and add the fixative.
23. Leave the gel for 10 min, then seal in a plastic bag.

4.2.2 Polymerase chain reaction

The polymerase chain reaction (PCR) is used to amplify small quantities of virus DNA and cDNA to concentrations detectable by gel electrophoresis,

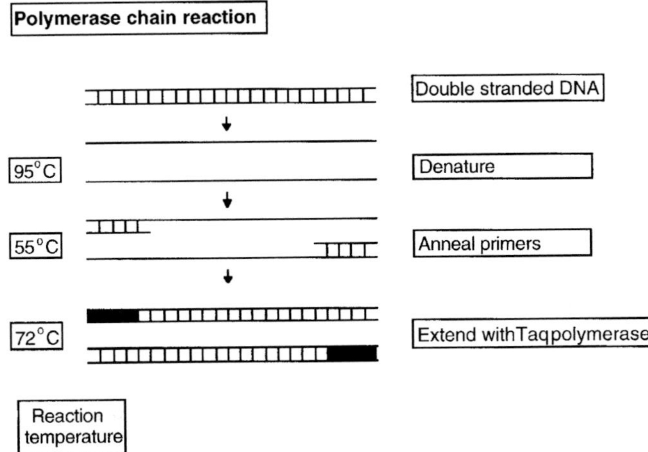

Figure 2. The polymerase chain reaction, in which two oligonucleotide primers, flanking the DNA fragment to be amplified, are annealed to denatured single-stranded DNA (ssDNA), and extended with Taq DNA polymerase. The reaction is performed at three different temperatures, and amplification results from repeat cycles at the three temperatures. One copy of target DNA can be amplified to over one million copies in less than 25 cycles.

through the use of thermostable DNA polymerases and oligonucleotide primer-initiated DNA synthesis. Three distinct events occur during a PCR cycle (*Figure 2*):

i. Denaturation at 92–96 °C of the extracted DNA or cDNA, often referred to as the template
ii. Annealing of the oligonucleotide primers to complementary sequences on the template at temperatures between 37 °C and 65 °C
iii. DNA synthesis by a thermostable polymerase through extension of the primers at 72–74 °C

Amplification by PCR requires the presence of two oligonucleotide primers 15–30 bases long (each complementary to sequences on one of the two strands of DNA and separated by several hundred base pairs), the four deoxynucleotide triphosphates (dNTPs), magnesium ions (usually as $MgCl_2$ in molar excess of the dNTPs), a thermostable DNA polymerase (e.g. Taq DNA polymerase or Pfu DNA polymerase), and template at picogram or nanogram quantities.

4.2.3 Nested PCR

Nested PCR, in which a second PCR is performed with primers internal to those used in the first PCR, has two advantages over a single-round PCR. The sensitivity of the assay is increased by more than a hundredfold, and the specificity is increased. A non-specific product amplified in the first round of a nested PCR would be unlikely to contain sequences complementary to the second set of primers used to initiate amplification in the second round.

Protocol 8. Nested PCR for detecting cytomegalovirus DNA

Equipment and reagents

- 10× PCR buffer (Life Technologies)
- 50 mM $MgCl_2$ (Life Technologies)
- dNTPs (10 mM mix) (Advanced Biotechnologies)
- Taq polymerase (5 U μl^{-1})(Life Technologies)
- Oligonucleotide primers (20 pmoles μl^{-1})
- Distilled water
- Template (extracted DNA: *Protocol 6*)
- 0.2 ml PCR tubes (Advanced Biotechnologies)
- Positive displacement pipettes (Gilson)
- Sterile filtered pipette tips (Anachem)
- Thermal cycler
- Nusieve 3:1 agarose (Flowgen)

Method

1. Prepare the first-round reaction mix for each specimen as follows: 10× PCR buffer II, 4.0 μl; 50 mM $MgCl_2$, 1.6 μl; dNTPs, 1.0 μl; Taq polymerase, 0.2 μl; CMV gB primer 1[a], 1.0 μl; CMV gB primer 2[b], 1.0 μl; distilled water, 31.2 μl.
2. Add 40 μl of first-round reaction mix (above) to 0.2 ml plastic PCR tubes.
3. Add 10 μl of the template to the reaction mix.
4. Load the tubes into the thermal cycler[e] and run the following programme: 94 °C for 2.5 min, followed by 30 cycles of: 95 °C for 30 s, 50 °C for 20 s, and 72 °C for 30 s, followed by holding at 4 °C.
5. Prepare the second-round PCR mix for each specimen as follows: 10× PCR buffer II, 2.0 μl; 50 mM $MgCl_2$, 0.8 μl; dNTPs, 0.5 μl; Taq polymerase, 0.2 μl; CMV gB primer 3[c], 1.0 μl; CMV gB primer 4[d], 1.0 μl; distilled water, 17.8 μl
6. Add 24 μl of the second-round PCR mix to a new PCR tube for each reaction.
7. Add 2 μl of the first-round product to the appropriate second-round reaction tube.
8. Load the tubes into the thermal cycler and run the following programme: 35 cycles of: 95 °C for 20 s, 50 °C for 20 s, 72 °C for 30 s, followed by holding at 4 °C.
9. After amplification, run nested PCR products on a 3% Nusieve 3:1 agarose gel. The size of the second-round product is predicted to be 297 bp.

CMV gB oligonucleotide primers:
[a] gB-1 5′ TGA GGA ATG TCA GCT TC 3′
[b] gB-2 5′ TCA TGA GGT CGT CCA GA 3′
[c] gB-3 5′ CCA GCC TCA AGA TCT TCA T 3′
[d] gB-4 5′ TCG TCC AGA CCC TTG AGG TA 3′
[e] Add molecular biology grade mineral oil to each tube if a thermal cycler without a heated cover is used.

Protocol 9. Agarose gel electrophoresis of PCR products

Equipment and reagents

- Nusieve 3:1 agarose (Flowgen)
- Microwave oven
- Gel tank (Anachem ORIGO midi gel [Anachem])
- TAE buffer (0.04 M Tris-acetate and 0.001 M EDTA)
- 10 mg ml^{-1} ethidium bromide
- TE buffer, pH 8.0 (10 mM Tris and 1 mM EDTA)
- Loading buffer (TE buffer containing 10% Ficoll [Sigma]and 0.25% Orange G [Sigma])
- UV transilluminator
- Polaroid camera

Method

1. Add 3 g of Nusieve 3:1 agarose to 100 ml of TAE buffer
2. Melt in the microwave at full power for 1–2 min.
3. Cool to 45°C, then pour into a gel plate fitted with a 16-slot comb.
4. Allow gel to set.
5. Add 20 μl of PCR product to 10 μl TE and 10 μl loading buffer in a microtitre tray.
6. Remove comb from the gel and add 40 μl of prepared product to the appropriate well.
7. Place the gel plate in the gel tank, and add TAE level with the gel.
8. Run the products into the gel for 5 min at a constant voltage of 150 V.
9. Flood the gel with TAE, making sure it is fully submerged.
10. Run at 150 V for 1.5 h.
11. Remove the gel plate from the electrophoresis tank.
12. Remove the gel from the plate, and stain in a tank with 300 ml TAE containing 100 μl of ethidium bromide.
13. Leave to stain for >30 min.
14. Examine on a UV transilluminator for bands of the predicted size.
15. Take a photographic record of the gel.

4.2.4 Reverse transcriptase PCR (RT-PCR)

A DNA copy (cDNA) of RNA virus genomes is obtained by means of reverse transcription (RT). Reverse transcriptases derived from avian myeloblastosis virus (AMV) or Moloney murine leukaemia virus (M-MuLV) have been used along with Tth DNA polymerase, which has reverse transcriptase activity. cDNA synthesis from virus RNA in the presence of reverse transcriptase is initiated by either a sequence-specific primer, or by random priming with hexameric oligonucleotides. Random priming offers the advantage of providing cDNA copies of all the RNA in the specimen. cDNA produced as a result of random priming can be used to detect any RNA virus genome in the PCR, provided virus-specific primers are available. Separate RT-reactions have to be performed for each RNA virus sought, if virus-specific primers are used for the initiation of cDNA synthesis.

Protocol 10. Reverse transcription with random primers

Equipment and reagents

- Random primers (Pharmacia)
- Extracted RNA
- Dry heating block at 70 °C (Grant)
- Dry heating block at 95 °C
- 10x PCR buffer II (Life Technologies)
- 50 mM $MgCl_2$ (Life Technologies)
- RNase-free distilled water
- 10 mM mix of dNTPs (Advanced Biotechnologies)
- 200 U μl^{-1} M-MLV reverse transcriptase (Life Technologies)
- 0.2 ml PCR tubes (Advanced Biotechnologies)
- Water bath at 37°C

Method

1. Add 1 μl of the random primer and 20 μl of the extracted RNA to a 0.2 ml PCR tube.
2. Incubate the tubes at 70 °C (ssRNA) or 96°C (dsRNA) for 5 min.
3. Chill the tubes on ice for 2 min.
4. Prepare the reverse transcription mix (RT-mix) for each specimen as follows: 10x PCR buffer II, 3.5 μl; 50 mM $MgCl_2$, 3.5 μl; dNTPs (10 mM mix), 0.7 μl; M-MLV reverse transcriptase, 1.0 μl; RNase-free distilled water, 5.3 μl.
5. Add 14 μl of the RT-mix to each tube.
6. Incubate the tubes at room temperature for 10 min.
7. Transfer the tubes to a waterbath and incubate at 37 °C for 1 h.
8. Transfer the tubes to a 95 °C heating block for 5 min.
9. Chill the tubes on ice for 2 min.
10. The cDNA can be used directly in the PCR, or stored at −20 °C for future use.

Protocol 11. Duplex (simultaneous) nested and semi-nested PCR to detect enteroviruses and echovirus type 22 respectively from random-primed cDNA

Equipment and reagents

- 10× PCR buffer II (Life Technologies)
- 50 mM $MgCl_2$ (Life Technologies)
- 10 mM mix of dNTPs (Advanced Biotechnologies)
- 5 U μl^{-1} Taq polymerase (Life Technologies)
- 20 pmoles μl^{-1} oligonucleotide primers
- Distilled water
- 0.2 ml PCR tubes (Advanced Biotechnologies)
- cDNA
- Thermal cycler

Method

1. Prepare the first-round reaction mix for each specimen as follows: 10× PCR buffer II, 4.0 μl; 50 mM $MgCl_2$, 1.2 μl; dNTPs (10 mM mix), 0.5 μl; Taq polymerase (5 U μl^{-1}), 0.2 μl; Enterovirus primer 1[a] (20 pmoles μl^{-1}), 1.0 μl; Enterovirus primer 2[b] (20 pmoles μl^{-1}), 1.0 μl; Echovirus 22 primer 1[c] (20 pmoles μl^{-1}), 1.0 μl; Echovirus 22 primer 3[d] (20 pmoles μl^{-1}), 1.0 μl; distilled water, 30.1 μl.
2. Add 40 μl of the first-round reaction mix to the 0.2 ml plastic PCR tubes.
3. Add 5 μl of the cDNA from the RT reaction to the first-round reaction mix.
4. Load the tubes into the thermal cycler and run the following programme: 94 °C for 4 min, followed by 30 cycles of: 95 °C for 30 s, 50 °C for 20 s, 72 °C for 30 s, followed by: 72 °C for 4 min. Hold at 4 °C.
5. Prepare the second-round PCR mix for each specimen as follows: 10× PCR buffer II, 2.5 μl; 50 mM $MgCl_2$, 1.0 μl; dNTPs (10 mM mix), 0.5 μl; Taq polymerase (5 U μl^{-1}), 0.2 μl; Enterovirus primer 3[e] (20 pmoles μl^{-1}), 1.0 μl; Enterovirus primer 4[f] (20 pmoles μl^{-1}), 1.0 μl; Echovirus 22 primer 1[c] (20 pmoles μl^{-1}), 1.0 μl; Echovirus 22 primer 2[g] (20 pmoles μl^{-1}), 1.0 μl; distilled water, 15.8 μl.
6. Add 24 μl of the second-round PCR mix to a new PCR tube for each reaction.
7. Add 2 μl of the first-round product to the appropriate second-round reaction tube.
8. Load the tubes into the thermal cycler and run the following programme: 35 cycles of: 94 °C for 20 s, 50 °C for 20 s, 72 °C for 30 s, followed by holding at 4°C.
9. After amplification, run the nested PCR products on a 3% Nusieve 3:1 agarose gel. (*Protocol 9*). The size of the second-round products is predicted to be 149 bp for enterovirus, and 171 bp for echovirus type 22.

Enterovirus and echovirus type 22 oligonucleotide primers:

[a] Enterovirus primer 1	5′ CGG CCC CTG AAT GCG GC 3′
[b] Enterovirus primer 2	5′ CAC CGG ATG GCC AAT CCA 3′
[c] Echovirus 22 primer 1	5′ CCA GCC GAA CAA CAT CTG 3′
[d] Echovirus 22 primer 3	5′ GGC CCA CTA GAC GTT TTT 3′
[e] Enterovirus primer 3	5′ CCC CTG AAT GCG GCT AAT 3′
[f] Enterovirus primer 4	5′ ATT GTC ACC ATA AGC AGC CA 3′
[g] Echovirus 22 primer 2	5′ ACC TTC TGG GCA TCC TTC 3′

4.3 Quantitative nucleic acid amplification techniques

The ability to quantify the number of copies of DNA or RNA in a clinical sample is crucial for monitoring the effectiveness of antiviral chemotherapy.

Thymidine kinase mutants of herpes simplex virus are selected in leukaemic patients being treated with acyclovir for HSV infections, and reverse transcriptase and protease mutants are selected during treatment with RT or protease inhibitors in patients infected with HIV. Also, the severity of symptoms associated with infection may directly correlate with virus load.

Although assays to amplify nucleic acids can be quantitative, through the use of serial dilutions or external standards with known quantities of DNA, they suffer from inbuilt errors. The rate at which target DNA accumulates is highly variable, leading to intra-assay variation, and is dependent on the reaction conditions and the amount of non-target DNA present in the reaction. The method of choice is the inclusion, in each reaction, of internal standards with the same amplification efficiency as the target. The internal standards are co-amplified in the same reaction as the target, so that for each amplification reaction the amount of specific product can be related to the amount of amplified internal control DNA. The internal standard may differ from the target by the inclusion of a restriction site, allowing the control to be differentiated from the target on the basis of size. Alternatively, part of the sequence of the control internal to the primers may be rearranged, allowing differentiation by the use of two oligonucleotide probes, one specific for the rearranged sequence, and the other for the target sequence (wild type).

The majority of quantitative assays used in diagnostic laboratories are sourced from commercial suppliers, due to the complexity of constructing, producing, and quality-controlling internal controls for quantitative nucleic acid amplification assays. Three commercial quantitative assays have found widespread use: quantitative PCR (Roche), NASBA (Organon Teknika) (*Figure 3*), and the bDNA assay (Chiron) (*Figure 4*).

5. Detection of specific anti-viral antibodies

The detection of specific anti-viral antibodies in serum, plasma, or saliva plays an important part in the diagnosis and epidemiology of virus infections. A wide range of assays is available for detecting virus-specific antibodies, and these include the complement-fixation test, haemagglutination inhibition tests, passive particle agglutination tests, indirect immunofluorescence (IF), radioimmunoassays, enzyme-linked immunosorbent assays (ELISA), and Western blotting. Also, the ability to differentiate between virus-specific antibodies of different immunoglobulin classes, IgM, IgA and, IgG, and to determine the avidity of the antibody detected, provides additional information which, in the absence of virus isolation, can help to provide a definitive diagnosis. IgM and IgA responses are transient, and often only detectable for 3–6 months after infection, whereas IgG is lifelong. The avidity of antibodies is low in the first 20–40 days after infection, and becomes high as the antibody response matures. Therefore, a serology laboratory should be equipped to perform a wide range

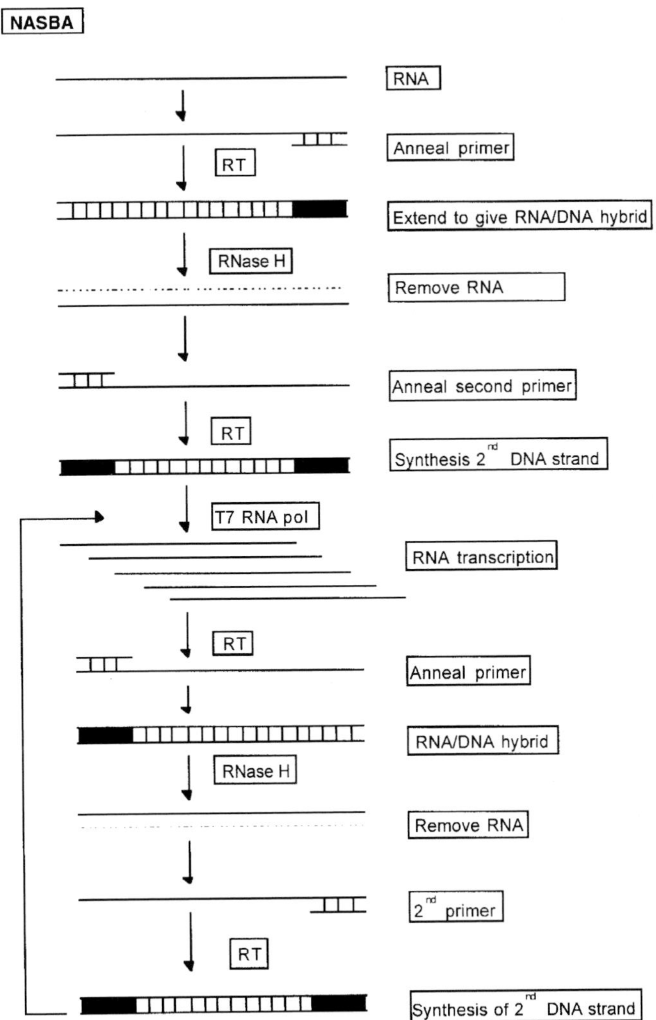

Figure 3. Nucleic acid sequence-based amplification (NASBA) amplifies target nucleic acid under isothermal conditions, through the use of three enzymes simultaneously. The reaction is initiated by the hybridization of one of the oligonucleotide primers, containing a T7 promoter sequence extension, to the target RNA. The primer is extended with reverse transcriptase (RT), and the RNA strand of the resulting RNA-DNA hybrid is degraded by RNase H. A second primer is annealed to the single stranded cDNA, and a second strand synthesized by RT, resulting in a dsDNA molecule with a T7 RNA promoter sequence. T7 RNA polymerase produces 100–1000 copies of RNA from each double-stranded DNA molecule. Amplification results from each synthesized RNA molecule serving as a template in the continued production of cDNA, and subsequent RNA synthesis.

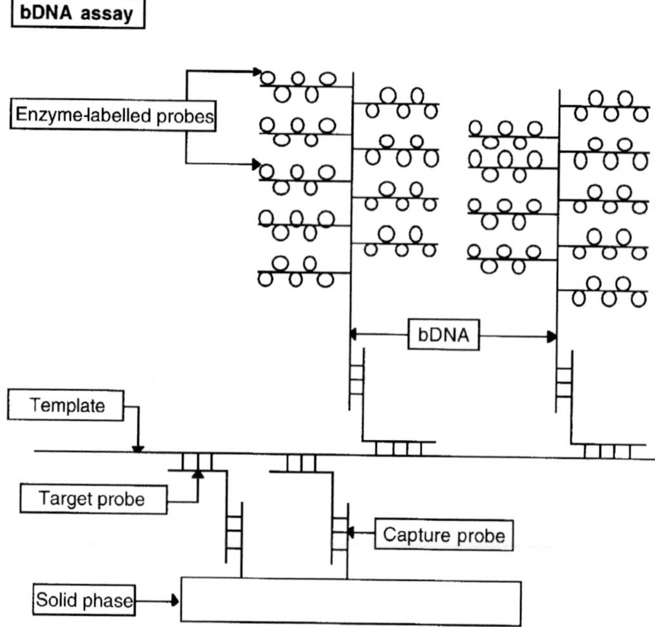

Figure 4. The bDNA assay is based on amplification of the detection signal rather than the target sequence, in contrast to PCR and NASBA. ssDNA or RNA is hybridized to multi-target probes that contain sequences complementary to the virus target, the capture probe, and the bDNA molecule. Probes and virus target are mixed, and DNA- or RNA-probe complexes are captured onto a solid phase. bDNA molecules are then hybridized, and enzyme-labelled probes, which catalyse dioxethane to give a chemiluminescent signal, are bound to the bDNA molecules.

of assays capable of measuring total antibodies, immunoglobulin class-specific antibodies, and determining antibody avidity.

5.1 Virus-specific IgM detection

The detection of virus-specific IgM plays an important role in the diagnosis of acute virus infections. IgM can be separated from other classes of immunoglobulins on a Sephadex G50 (Pharmacia) column, or by ultracentrifugation on a sucrose gradient, as a result of their greater molecular weight. These procedures are labour-intensive, and have largely been superseded by class-specific immunoglobulin assays, such as μ-capture ELISAs for detecting virus-specific IgM.

5.1.1 Antibody-capture ELISA

IgM-capture (μ-capture) ELISAs use anti-human IgM, bound to a solid phase, to capture IgM molecules from the sample. All other classes of immunoglobulin are unbound, and are removed during washing. Virus-specific IgM is then

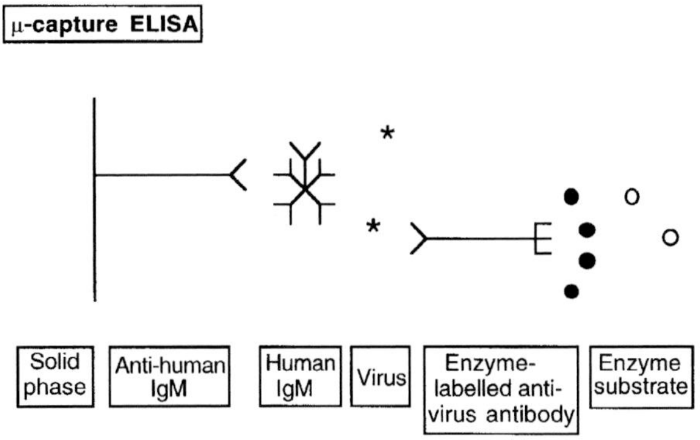

μ-capture ELISA

Solid phase | Anti-human IgM | Human IgM | Virus | Enzyme-labelled anti-virus antibody | Enzyme substrate

Figure 5. IgM capture ELISA (μ-capture) for detecting virus-specific IgM. Anti-IgM, immobilized on the solid phase, captures IgM from the patient's serum. The virus-specific IgM binds virus antigen, which is detected with an enzyme-labelled anti-virus antibody. The bound enzyme catalyses the enzyme substrate to produce a colour reaction.

detected by adding a specific virus antigen followed by an enzyme-labelled specific anti-viral antibody. The bound enzyme, in the presence of an appropriate substrate, will catalyse the production of a coloured product (*Figure 5*).

Protocol 12. IgM antibody-capture ELISA (μ-capture)

Equipment and reagents

- Flat-bottomed microtitre plates (Becton Dickinson)
- Rabbit-anti-human IgM (μ-chain specific) (Dako)
- PBS, pH 7.2 (Oxoid)
- Tween 20 (Sigma)
- Microtitre plate washer (Denley)
- Bovine serum albumin (fraction V) (Sigma)
- Skimmed milk powder (Oxoid)
- Normal rabbit serum (Sigma)
- Test serum
- Virus antigen
- Fetal bovine serum (Sigma)
- Incubator at 37°C
- HRPO-labelled anti-virus antibodies (Dako)
- Tetramethylbenzidine (Sigma)
- Hydrogen peroxide (Sigma)
- Substrate buffer (pH 6.0) (*Protocol 4*)
- Spectrophotometer with 450 nm wavelength filter.
- 2 M sulfuric acid

Method

1. Coat the wells of a microtitre plate with 100 μl of antibody to human IgM (μ-chain) at a predetermined concentration in carbonate-bicarbonate buffer, pH 9.6 (*Protocol 4*).
2. Cover the plate, and incubate overnight at 4 °C.
3. Wash the plate three times with PBS-Tween (*Protocol 4*).
4. Add 200 μl of PBS containing either 0.5% bovine serum albumin (BSA), 5% normal rabbit serum (NRS), or 5% skimmed milk powder

Protocol 12. *Continued*

(SMP), and incubate the plate at 37°C for 30 min to block the uncoated sites.

5. Remove the blocking agent by aspiration.
6. Add 100 μl of the test serum diluted 1 in 100 in PBS-Tween (include positive and negative controls).
7. Incubate the plate for 2 h in a moist atmosphere at 37 °C.
8. Wash the plate five times with PBS-Tween as before.
9. Add 100 μl of the virus antigen, diluted in PBS-Tween containing 10% fetal bovine serum.
10. Incubate the plate overnight in a moist atmosphere at room temperature.
11. Wash the plate five times in PBS-Tween as before.
12. Add 100 μl of the enzyme-labelled anti-virus antibody, diluted in PBS-Tween to a pre-determined dilution.
13. Incubate the plate for 2 h in a moist atmosphere at 37 °C.
14. Wash the plate five times with PBS-Tween as before.
15. Add 100 μl of the tetramethylbenzidine-H_2O_2 in substrate buffer to each well (*Protocol 4*).
16. Incubate the plate for 30 min at room temperature.
17. Add 100 μl of the 2 M sulfuric acid to stop the reaction, and measure the optical density at a wavelength of 450 nm.

5.2 Detection of virus-specific IgG

IgG responses to infection occur later than IgM and IgA responses. Therefore, it is possible to diagnose an acute infection by demonstrating a seroconversion or significant rise in IgG levels in acute and convalescent samples. In practice, paired sera are not always obtained, and in some infections such as with parvovirus B19, virus-specific IgG is often already detectable when the patient presents with symptoms. IgG assays are more often used to determine a patient's immune status prior to, or after, vaccination, or to determine the prevalence of a virus in the population by identifying individuals with pre-existing virus-specific antibodies.

5.2.1 ELISAs for detecting virus-specific IgG

IgG antibody-capture (γ-capture), indirect, and competitive ELISAs are all used to detect virus-specific IgG. In the indirect ELISA the presence of IgG,

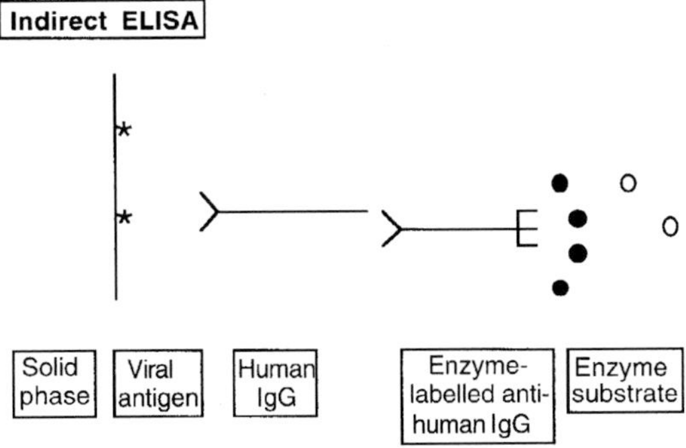

Figure 6. Indirect ELISA for detecting virus-specific IgG. Virus antigen, bound to the solid phase, captures virus-specific antibodies. Virus-antibody complexes are detected through the use of an enzyme-labelled anti-human IgG antibody.

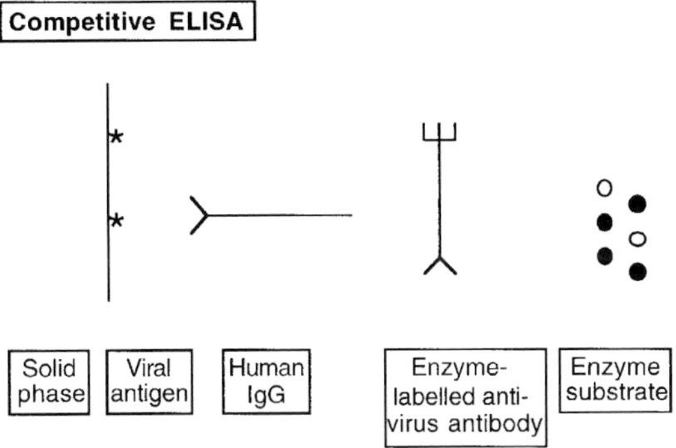

Figure 7. Competitive ELISA for detecting virus-specific IgG. Virus-specific antibody, captured through its interaction with virus antigen bound to the solid phase, prevents the binding of an enzyme-labelled anti-virus antibody. Therefore, in this assay a colour reaction represents a negative result.

complexed with an antigen bound to the solid phase, is detected with an enzyme-labelled anti-human IgG antibody (*Figure 6*). In the competitive ELISA a coloured product is not produced, since specific IgG, present in the patient's serum, prevents a labelled specific antibody from binding to the antigen (*Figure 7*). IgG antibody-capture and indirect ELISAs can be used to quantify virus-specific IgG, whereas competitive assays will only determine the presence or absence of antibody, and are therefore more suited to epidemiological or antibody prevalence studies. For further details on ELISAs see Gray and Wreghitt (15).

Protocol 13. Indirect ELISA for detecting virus-specific IgG

Equipment and reagents

- Flat-bottomed microtitre plates (Becton Dickinson)
- Virus antigen
- PBS (Oxoid)
- Tween 20 (Sigma)
- Refrigerator at 4 °C
- Microtitre plate washer (Denley)
- Skimmed milk powder (Oxoid)
- Bovine serum albumin (Sigma)
- Normal rabbit serum (Sigma)
- Test serum
- Incubator at 37 °C
- HRPO-labelled anti-human IgG (Dako)
- Tetramethylbenzidine (Sigma)
- Hydrogen peroxide (Sigma)
- Substrate buffer (pH 6.0) (*Protocol 4*)
- 2 M sulfuric acid
- Spectrophotometer with 450 nm wavelength filter

Method

1. Coat the wells of a microtitre plate with 100 μl of the virus antigen at a predetermined concentration (and alternate wells with uninfected control antigen where appropriate) in carbonate-bicarbonate buffer, pH 9.6 (*Protocol 4*).
2. Cover the plate and incubate overnight at 4 °C in a moist atmosphere.
3. Wash the plate three times with PBS-Tween (*Protocol 4*)
4. Add 200 μl of PBS containing either 0.5% BSA, 5% NRS, or 5% SMP, and incubate at 37 °C for 30 min to block uncoated sites.
5. Remove the blocking agent by aspiration.
6. Add 100 μl of the test serum, diluted 1 in 100 in PBS-Tween (include positive and negative controls in each assay run).
7. Incubate the plate at 37 °C in a moist atmosphere for 1 h.
8. Wash the plate five times with PBS-Tween as before.
9. Add 100 μl of the enzyme-labelled anti-human IgG in PBS-Tween at a predetermined dilution (usually between 1 in 1000 and 1 in 5000).
10. Incubate the plate at 37 °C in a moist atmosphere for 1 h.
11. Wash the plate five times with PBS-Tween as before.
12. Add 100 μl of tetramethylbenzidine-H_2O_2 in substrate buffer (*Protocol 4*).
13. Incubate the plate at room temperature for 30 min.
14. Add 100 μl of 2 M sulfuric acid to stop the reaction, and measure the optical density at a wavelength of 450 nm.

5.2.2 Antibody avidity determination

The early virus-specific IgG response after primary infection is made up primarily of low avidity antibodies, with avidity increasing in a few weeks to months after infection (16). The ability to differentiate between the presence of low avidity and high avidity antibodies in serum by the incorporation of mild denaturing agents, such as 8 M urea or 20–100 mM diethylamine, into antibody assays, to reduce the binding of low avidity IgG, helps to discriminate between recent or primary, and past or secondary infections, in the absence of paired serum samples (17). Both indirect immunofluorescence assays and ELISAs have been successfully modified for antibody avidity determination, by the addition of 8 M urea to the wash steps subsequent to the incubation of the patient's serum. Assays are performed with and without the addition of urea. If the titre, in the case of indirect IF assays is reduced by more than four-fold, or the optical density in ELISAs by more than 50% in the presence of urea, then the antibody is of low avidity.

Protocol 14. Determining antibody avidity by ELISA

Equipment and reagents

- Flat-bottomed microtitre plates (Becton Dickinson)
- Virus antigen
- PBS (Oxoid)
- Tween 20 (Sigma)
- 8M urea (Sigma)
- Refrigerator at 4 °C
- Microtitre plate washer (Denley)
- Skimmed milk powder (Oxoid)
- Bovine serum albumin (Sigma)
- Normal rabbit serum (Sigma)
- Test serum
- Incubator at 37 °C
- HRPO-labelled ant-human IgG (Dako)
- Tetramethylbenzidine (Sigma)
- Hydrogen peroxide (Sigma)
- Substrate buffer (pH 6.0)
- 2 M sulfuric acid
- Spectrophotometer with 450 nm wavelength filter

Method

1. Coat the wells of a microtitre plate with 100 μl of virus antigen at a predetermined concentration (and alternate wells with uninfected control antigen where appropriate) in carbonate-bicarbonate buffer, pH 9.6 (*Protocol 4*).
2. Cover the plate and incubate overnight at 4 °C in a moist atmosphere.
3. Wash the plate three times with PBS-Tween (*Protocol 4*)
4. Add 200 μl of PBS containing either 0.5% BSA, 5% NRS, or 5% SMP, and incubate at 37 °C for 30 min to block the uncoated sites.
5. Remove the blocking agent by aspiration.
6. Add 100 μl of the test serum diluted 1 in 100 in PBS-Tween to each of two wells (include positive and negative controls in each assay).
7. Incubate the plate at 37 °C in a moist atmosphere for 1 h.
8. Wash the urea-designated wells three times with PBS containing 8M urea. After filling the wells, leave for 1 min before aspirating the wash fluid.
9. Wash the controls wells three times with PBS-Tween, allowing a 1 min soak after each wash.

Protocol 14. *Continued*

10. Wash all the wells three times with PBS-Tween. Do not include a soak time.
11. Add 100 μl of enzyme-labelled anti-human IgG in PBS-Tween at a predetermined dilution (usually between 1 in 1000 and 1 in 5000).
12. Incubate the plate at 37°C in a moist atmosphere for 1 h.
13. Wash the plate five times with PBS-Tween as before.
14. Add 100 μl of tetramethylbenzidine-H_2O_2 in substrate buffer (*Protocol 4*).
15. Incubate the plate at room temperature for 30 min.
16. Add 100 μl of 2 M sulfuric acid to each well to stop the reaction, and measure the optical density at a wavelength of 450 nm. Calculate the reduction of optical density values in the presence of urea, and determine the antibody avidity.

References

1. Schmidt, N. J., and Emmons, R. W. (1989). *Diagnostic procedures for viral, rickettsial and chlamydial infections*, p. 19. APHA, Washington.
2. Gray, J. J., and Brenwald, N. P. (1991). *J. Virol. Meth.*, **32**, 163.
3. Bradstreet, C. M. P., and Taylor, C. E. D. (1962). *Monthly bulletin of the Ministry of Health, Public Health Laboratory Service*, **21**, 96.
4. Griffiths, P. D., Panjwani, D. D., Stirk, P. R., Ball, M. G., Ganczakowski, M., Blacklock, H. A., and Prentice, H. G. (1984). *Lancet*, **ii**, 1242.
5. Gray, J. J., Major, A., Kewley, D. R., and Coombs, R. R. A. (1990). *Serodiagnosis and Immunotherapy in Infectious Diseases*, **4**, 143.
6. Simmonds, P. (1995). In *Medical virology. a practical approach* (ed. U. Desselberger), p. 107. Oxford University Press, Oxford.
7. Guatelli, J. C., Whitfield, K. M., Kwoh, D. Y., Barringer, K. J., Richman, D. D., and Gingeras, T.R. (1990). *Proc. Nat. Acad. Sci. USA*, **87**, 1874.
8. Urdea, M. S., Horn, T., Fultz, T. J., Anderson, M., Running, J. A., Hamren, S., Ahle, D., and Chang, C. A. (1991). *Nucleic Acid Res. Symp. Ser.*, **24**, 197.
9. Sambrook, J., Fritsch, E. F., and Maniatis, T. (1989). *Molecular cloning. a laboratory manual*, 2nd edn. Cold Spring Harbor Laboratory Press, NY.
10. Boom, R., Sol, C. J. A., Salimans, M. M. M., Jansen, C. L., Wertheim-van Dillen, P. M. E., and van der Noordaa, J. (1990). *J. Clin. Microbiol.*, **28**, 495.
11. Desselberger, U. (1989). In *Viruses and the gut* (ed. M. J. G. Farthing), p. 55. Swan Press, London.
12. Lonsdale, D. M. (1979). *Lancet*, **I**, 849.
13. Huang, E., Alford, C. A., Reynolds, D. W., Stagno, S., Pass, R. F. (1980). *New Eng. J. Med.*, **303**, 958.
14. O'Donnell, B., Bell, E., Payne, S. B., Mautner, V., and Desselberger, U. (1986). *J. Med. Virol.*, **18**, 213.

15. Gray, J. J., and Wreghitt, T. G. (1995). In *Medical virology. a practical approach* (ed. U. Desselberger), p. 33. Oxford University Press, Oxford.
16. Inouye, S., Hasegawa, A., Matsuno, S., and Katwo, S. (1984). *J. Clin. Microbiol.*, **20**, 525.
17. Gray, J. J. (1995). *J. Virol. Meth.*, **52**, 95.

5

Electron microscopy of viruses

S. S. BIEL and H.R. GELDERBLOM

1. Introduction

Though occasionally considered as laborious, old-fashioned, and unnecessary, the detection and identification of viruses and other agents by electron microscopy (EM) can make an important contribution to many basic and applied biomedical routine procedures. In some fields, for example, vaccine production, EM even forms an integral part of the production controls (*Table 1*). While the majority of direct and indirect virus detection techniques (e.g. ELISA, PCR) are highly sensitive, but also highly specific for just one individual species or family of agents, EM works differently. Based on its undirected approach and its high resolution, EM is able to detect routinely and characterize morphologically all structures greater than 15 nm in diameter, which fortunately coincides with the size of the smallest viruses (1).

More than 3600 animal, plant, and bacterial virus isolates have been described, with the majority of them being classified according to particle morphology and genome properties as members of one of the 73 established virus families (1). Using morphological criteria, that is, assessing size, overall morphology, and substructure, EM permits a rapid and direct assignment of a

Table 1. Fields of application of EM in cell-culture specimens

Application	Examples
Characterization of cell lines	Description of morphology, adhesion properties, contamination.
Assessment of biological purity	Evaluation of master seed cell cultures in vaccine production; evaluation of vaccines (see *Figure 4a*).
Test for virus contamination	Search for e.g. polyomaviruses (SV 40), paramyxoviruses (SV 5), retroviruses (HERV) (see *Figures 1 & 4*).
Test for other adventitious agents	Search for bacteria, fungi, mycoplasms.
Clinical virology	Examination of diagnostic cell cultures for the presence of pathogenic agents (see *Figures 1, 4 & 5*).

Dedicated to Hermann Frank, Tübingen – on occasion of his 75th birthday – who introduced me into the art and science of electron microscopy (hg).

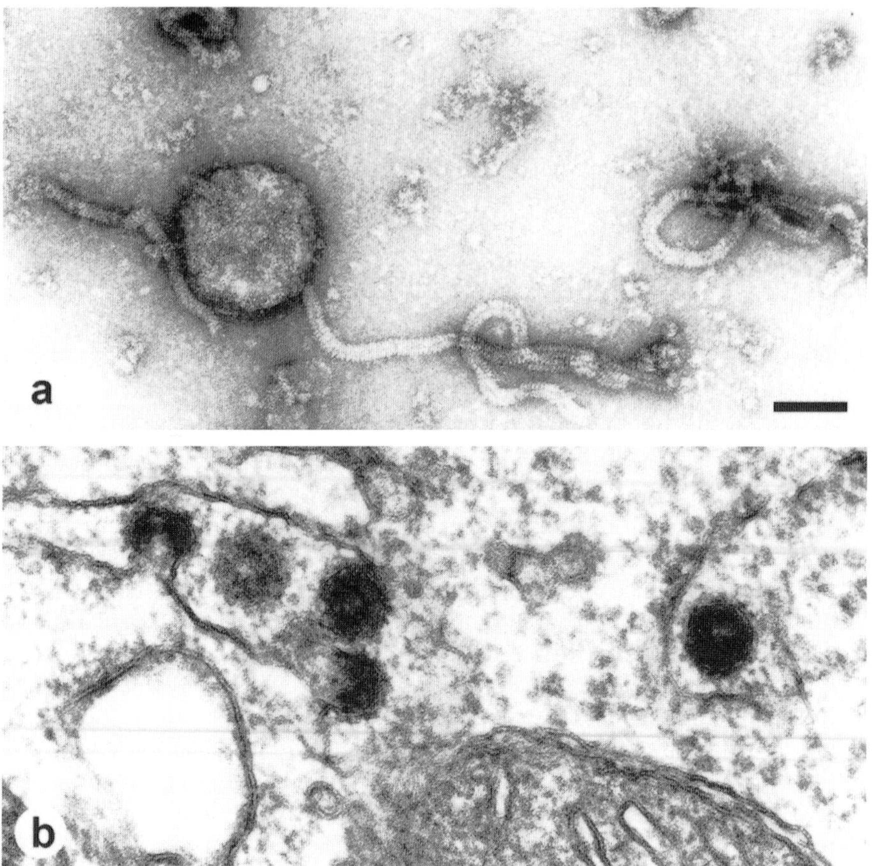

Figure 1 'Common' virus contaminants in cell culture. (a) Uranyl acetate negative staining (two-step, 2% UAc) of supernatants of a permanent cell culture, showing a single intact enveloped particle not penetrated by the stain, as well as free RNP with 'herring-bone' structure and a free central channel, typical of members of the paramyxovirus family. (b) The ultrathin section of a myeloma cell reveals endogenous retrovirus particles budding into and free in the endoplasmic reticulum. Scale bar = 100 nm.

detected particle to its specific virus family. Based on the principle of the 'open view', EM is also able to detect multiple infections and even infections that have not been screened for because they had been considered unlikely to occur.

The aim of this chapter is to illustrate the possibilities and limitations of EM in the evaluation of cell culture specimens, with the main emphasis on basic techniques applicable in fields from classic virus diagnosis to good manufacturing practice (GMP). We will concentrate on practical aspects, being aware, however, that proper specimen preparation and evaluation are learned best by practice in the laboratory of an experienced electron microscopist. Detailed instructions for the observation and recording of images using any particular instrument will not be given here. Similarly, complex techniques such as diffraction methods and image reconstruction are intentionally not covered.

For virus detection, conventional transmission electron microscopy (TEM), rather than scanning electron microscopy (SEM), is generally preferable. TEM can present both external and internal virus structures as 2D information, while SEM information is confined to surface topology presented in 3D. In TEM, a coherent beam of electrons passes through specimen, interacting with individual components. Hitting atomic nuclei, electrons are deflected at wide angles without loss of energy (elastic scattering), while interaction with the atomic shell results in energy loss and low angle scattering, that is, inelastic deflection of the electrons. Both events, elastic and inelastic scattering, contribute to image formation in the TEM (2, 3, 4).

The electrons in an EM beam are monochromatic, conventionally accelerated to 60–150 keV, and able to visualize properly specimens with a thickness of up to 100 nm. While thin specimens allow electrons to make singular encounters with the specimen's atoms during their passage, in thicker specimens electrons will interact significantly with several atoms, resulting in two effects:

- electrons become multiply scattered and deflected out of the image plane, that is, the image becomes 'dark' and blurred.
- the electrons after passage show a wide range of energy losses, resulting in wavelength differences (chromatic aberration).

Such electrons no longer focus properly into one image plane, and thus give rise to a blurred image. To attain clear, high contrast, and medium to high resolution TEM information, chromatic aberration and multiple scattering have to be reduced, for example, by using relatively thin specimens. Both material contrast, that is, the scattering of electrons out of the image, and phase contrast, the result of the interaction of different electrons, contribute to the formation of the image. There are two principal techniques for preparing virus specimens from cell culture for TEM, negative staining and thin sectioning, which together fulfil the great majority of demands (*Figures 2 and 3*).

Alternative preparation techniques, for example, heavy-metal shadowing, and replica and freeze-preparation techniques (freeze-fracturing, cryo-EM) are in part research techniques, that is, laborious and technically demanding, and the results often depend on the purity and concentration of the specimen. A description of these additional techniques goes beyond the scope of this chapter. Here we describe in detail a limited number of methods useful in routine virus cell-culture TEM, aiming to give practical advice and hints in particular to the beginner, and pointing to different strategies and methods in specimen preparation and evaluation. The advanced reader interested in further details, for example, in virology, image formation, and more sophisticated preparation techniques is referred to the literature listed at the end of this chapter.

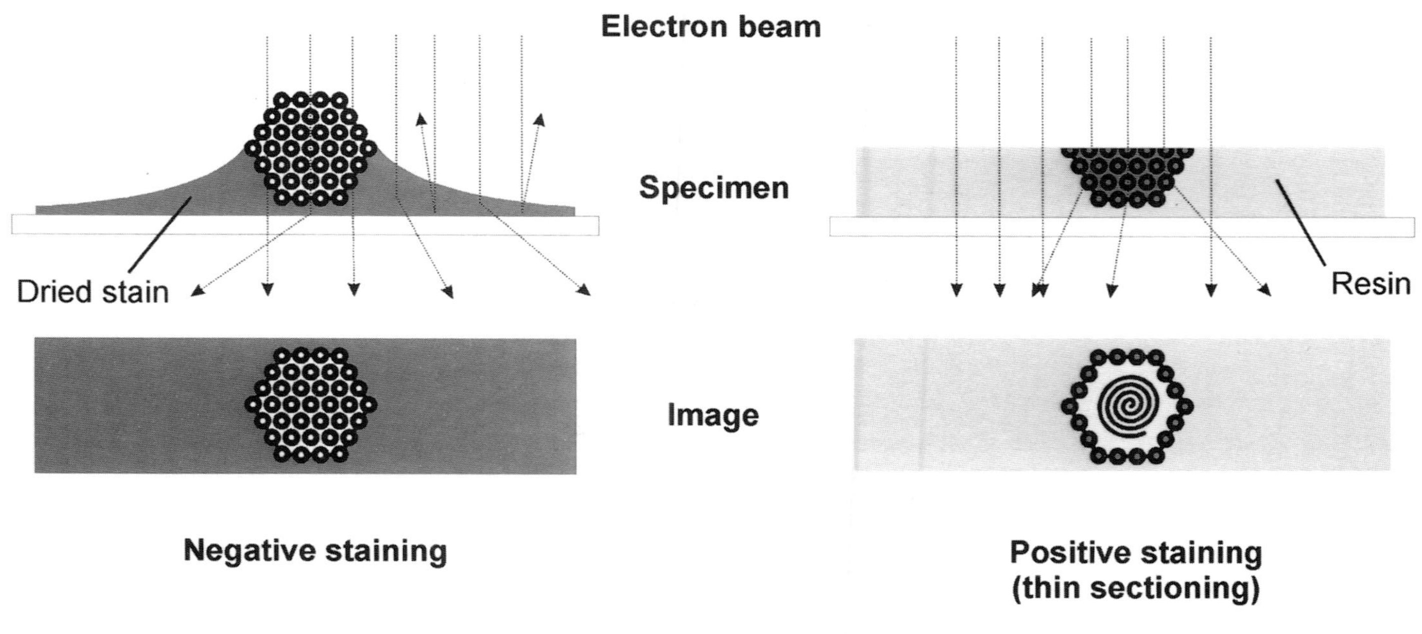

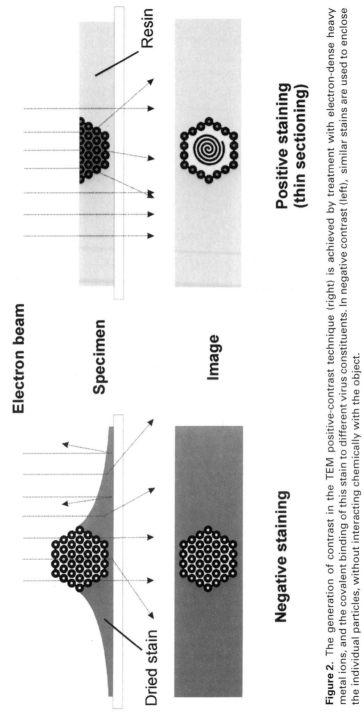

Figure 2. The generation of contrast in the TEM positive-contrast technique (right) is achieved by treatment with electron-dense heavy metal ions, and the covalent binding of this stain to different virus constituents. In negative contrast (left), similar stains are used to enclose the individual particles, without interacting chemically with the object.

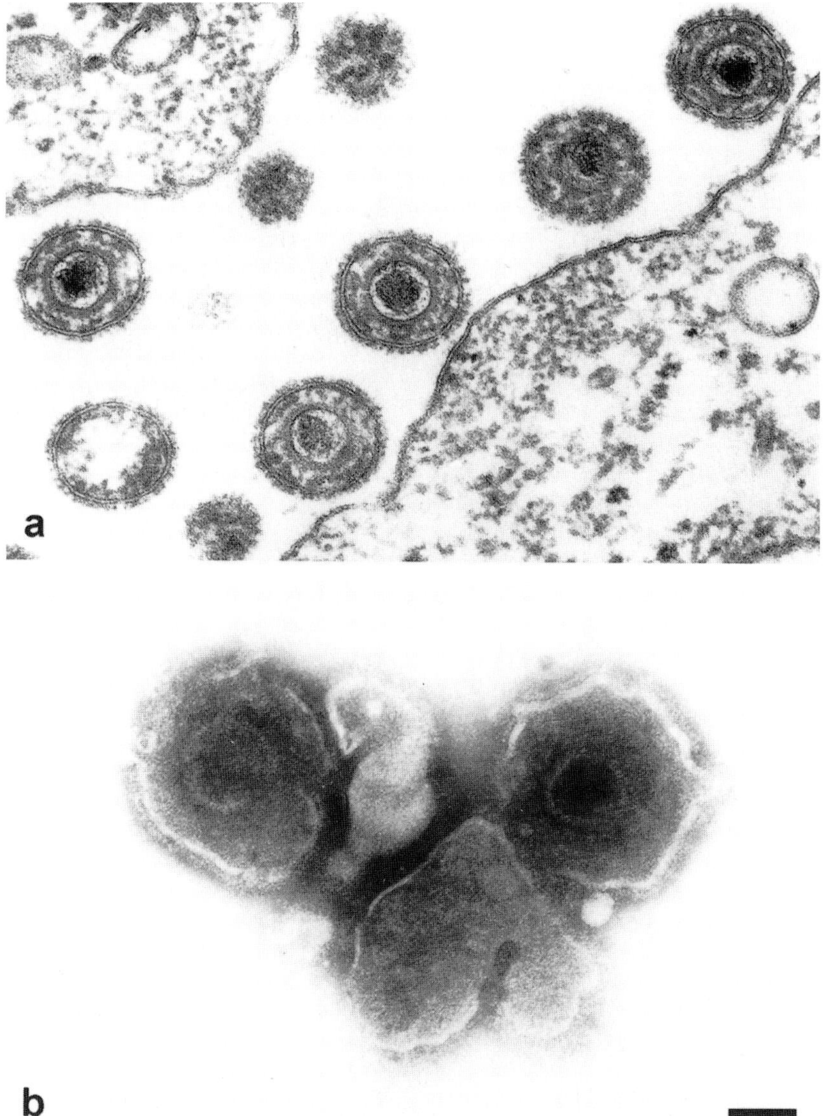

Figure 3. Comparison of positive and negative contrast effects in TEM. (a) Thin section TEM of herpesvirus shows clearly the virus genome contained in the centrally located, electron-dense toroid, the tegument surrounding the isometric capsid, and the limiting outer membrane studded with glycoprotein projections. (b) After two-step negative staining with 2% UAc of cell culture-derived herpesviruses inner virus constituents, the capsomeres, for example, are difficult to detect, while the glycoprotein projections on the outer membrane are shown more clearly. Scale bar = 100 nm.

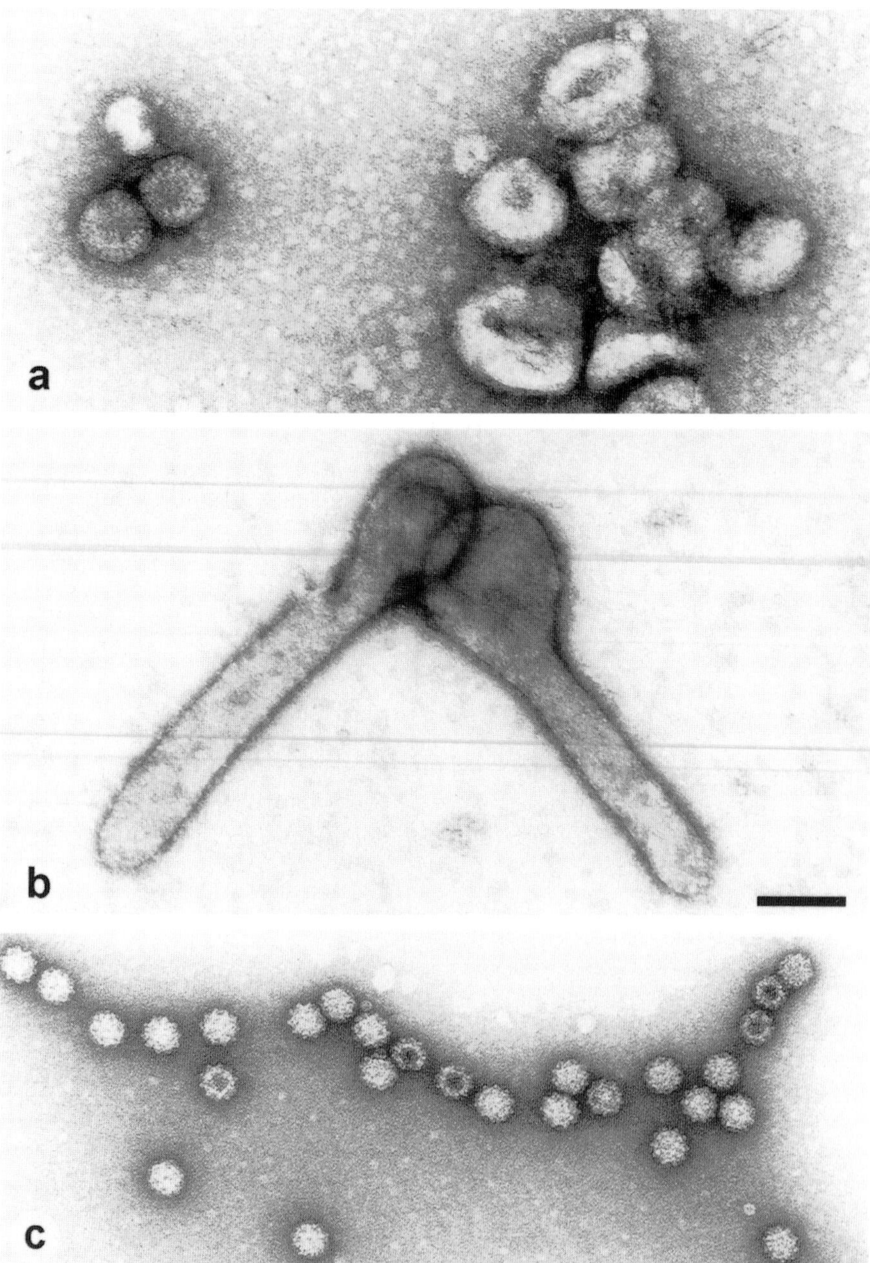

Figure 4. Virus morphology: a survey. (a) Rotaviruses (naked capsid, 70 nm diameter) and coronaviruses (enveloped particles, about 120 nm diameter), derived from a vaccine against cattle diarrhoea. (b) Marburg virus, enveloped, with small surface projections, 80 × 500 nm. (c) Rabbit haemorrhagic disease calicivirus (small round structured

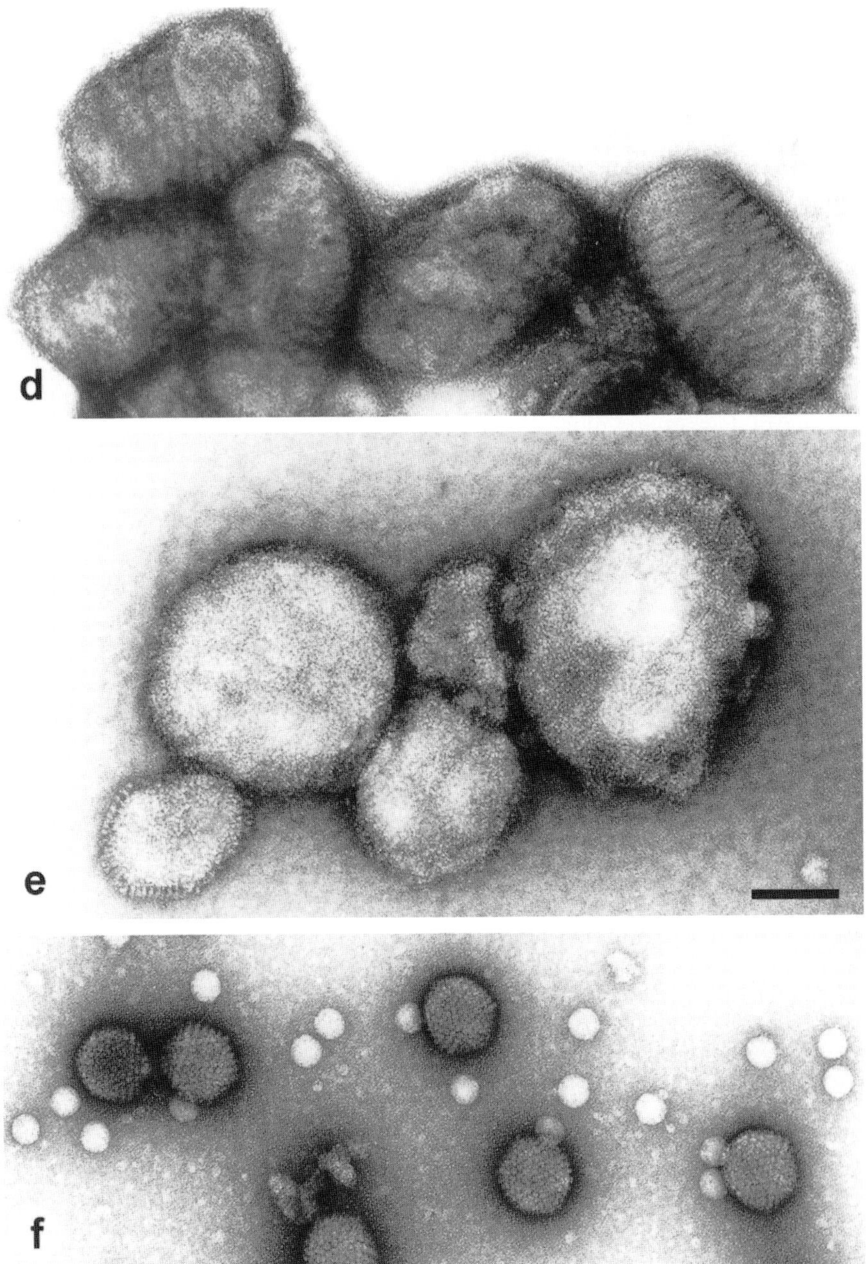

virus SRSV, 27–30 nm in diameter). (d) Parapox viruses isolated from seal, brick-shaped, 150 × 250 nm. (e) Paramyxovirus (Newcastle disease virus; varying size classes, envelope projections). (f) Adenovirus (diameter 79 nm) and adeno-associated virus (diameter 20–25 nm). (a) – (f) were two-step negatively stained with 1% UAc. Scale bar = 100 nm.

2. Virus structure and strategies for virus detection

While small viruses (e.g. parvoviruses and picornaviruses) contain the nucleic acid genome in a regular protein shell made up of multiple copies of one or a few proteins forming the capsid (*Figure 4c*), bigger and more complex viruses show further morphological substructures, for example, an additional lipid membrane studded with glycoprotein projections (*Figure 3*). Size, shape, and substructure differ significantly between the different virus families. Based on these structural peculiarities (*Figure 4*), virus differentiation by EM is feasible (1, 5, 6, 7, 8).

Virus-encoded biochemical constituents (e.g. polypeptides, nucleic acids) assemble into morphological subunits (e.g. capsomeres, ribonucleoprotein – RNP). The architecture of virus capsids follows either helical or icosahedral symmetry. Bigger and more complex viruses often show combinations of both virus symmetries with their different constituents. The determination of the size of an unknown particle, its capsid symmetry, the presence of additional structural components (e.g. glycoprotein projections, lipid membranes) finally leads to the assignment of the particle to a particular virus family.

Using negative staining and thin section TEM, virus particles and their constituents are presented in different ways. In negative staining, virus surface structures, that is, the capsid and its capsomeres, or the lipid membrane, are outlined precisely with high resolution by the electron-dense stain. Inner constituents, for example, a secondary capsid, core, or ribonucleoprotein, are not easily accessible. In thin sections, all virus components are detected, based on their different chemical affinities during specimen treatment (*Figure 3*).

Though the structural detail recognized in thin sections is not as high as in negative staining (4 nm versus 2 nm in routine TEM), sections often contain more structural information on a particular virus. Therefore, the more laborious thin sectioning can be particularly rewarding, especially in the search for virus in diagnostic cell cultures (*Figure 1, 5–6*).

3. Specimen preparation

Before analysing virus and cell culture specimens by TEM, a series of preparation steps have to be performed. Although the preparation for negative staining and thin section TEM differs, both techniques benefit from concentration steps. This is due to the fact that TEM is investigating very small volumes – though at high magnification – and therefore requires high absolute particle densities for efficient detection. Using negative staining, the removal of cellular debris, as well as concentration, facilitate efficient EM analysis. However, in thin-section EM, the cells and/or cellular components are often embedded as a pellet together with the virus, thus preserving the actual state of the virus–cell interactions. Embedding and thin sectioning of ultracentrifuge

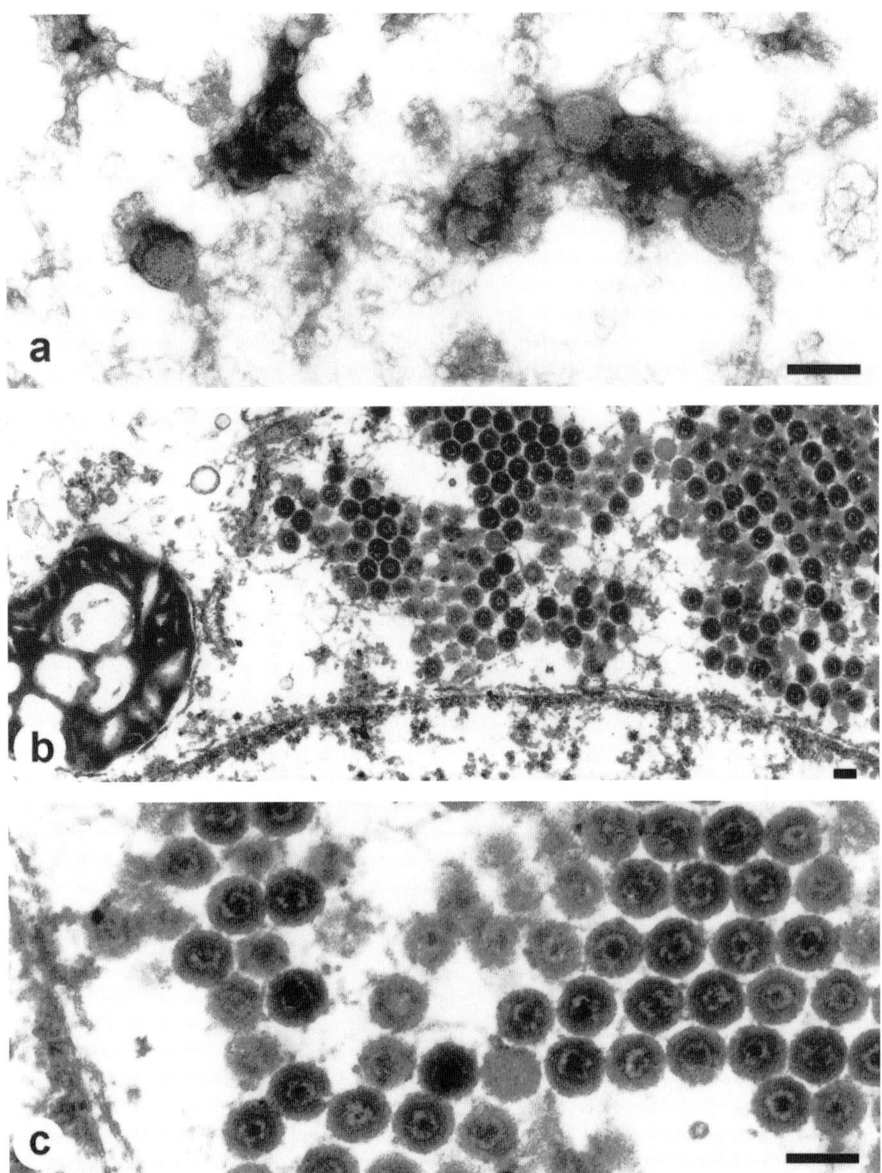

Figure 5. Parallel application of negative staining and thin section TEM on a diagnostic cell culture. (a) After two-step negative staining with 2% UAc, the presence of suspicious particles was barely recognizable in the culture supernatant, due to the high degree of contamination. Diagnosis remained uncertain until thin section TEM was performed. (b), (c) Ultrathin sections revealed virus crystals in the cytoplasm, composed of particles which, according to size, fine structure, and localization can be assigned to the reovirus family. (b) survey magnification, (c) particles at a higher magnification. Scale bars = 100 nm.

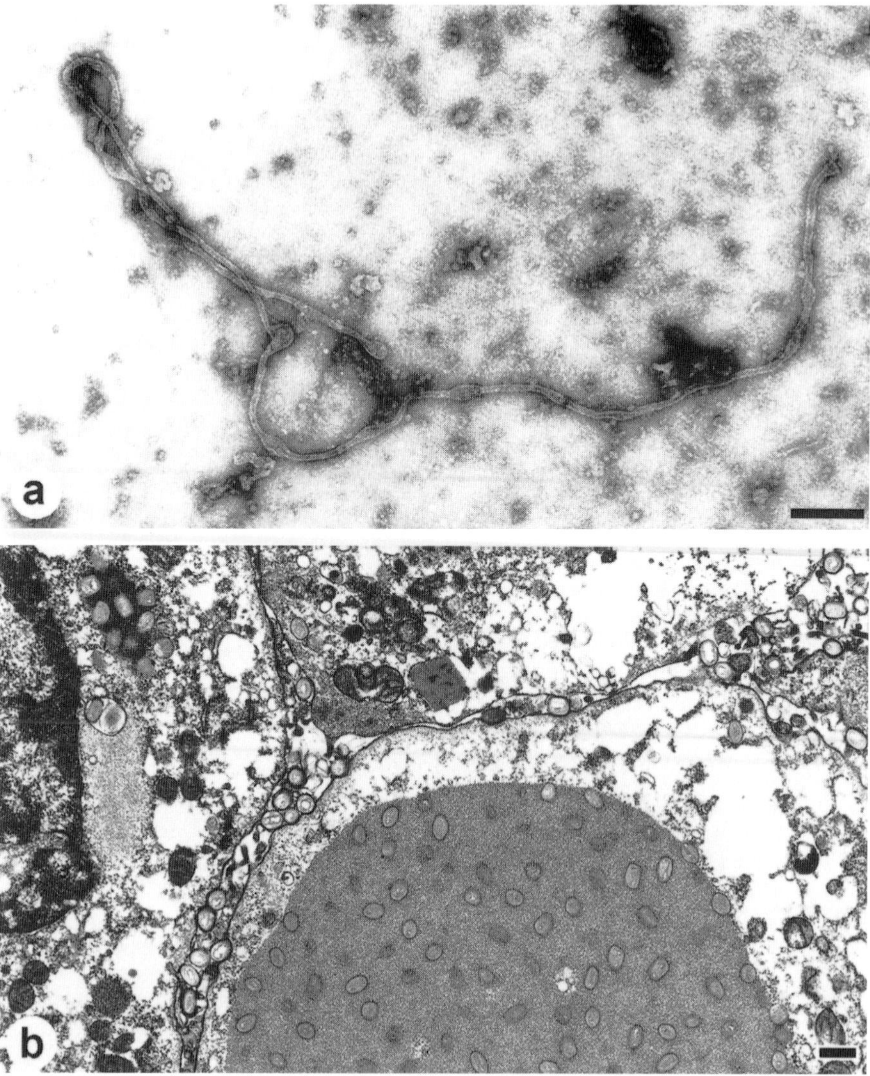

Figure 6. Comparative TEM on a diagnostic cell culture inoculated with liver homogenate from a diseased elephant. (a) After two-step negative staining with 2% UAc of the culture supernatants, these membrane-bound tubular structures, reminiscent of filoviruses, were found. The tubules probably represent parts of the endoplasmic reticulum derived from cells suffering from severe cytopathic effects and undergoing destruction. (b) Thin section TEM showed clearly cytoplasmic inclusions of poxviruses and their inclusion bodies. Scale bars = 250 nm.

Table 2. Techniques for virus enrichment and purification

Method	Enrichment	Comments
Low-speed centrifugation	Zero	Essential to remove cellular debris.
Ultracentrifugation	10–500 fold	Commonly used.
Beckman Airfuge®	50 fold	Enrichment from small volumes.
Direct onto grid sedimentation	50–200 fold	Easy to use (e.g. 100 000 *g*, 30–60 min).
Gradient centrifugation	100–500 fold	Very efficient but time consuming.
Sucrose cushion centrifugation	20–200 fold	Time consuming; preserves labile virus structures.
Ultrafiltration: Centricon®	10–100 fold	Expensive; not suitable for labile viruses.
Agar filtration	10 fold	Useful before negative staining.
Molecular sieve chromatography	Dilution!	Efficient purification.
Precipitation	$>$ 100 fold	Co-precipitation of contaminants.
Bioaffinity techniques	10–500 fold	Specific antibodies or ligands are required.
Solid-phase immuno-EM	10–500 fold	Difficult to establish.
Affinity chromatography	10–100 fold	Rarely used.

sediments from cell culture supernatants or cell sediments can be particularly useful in virus diagnosis.

Depending on the virus–cell system and the development of virus cytopathic effect (CPE), viruses are released efficiently into the supernatant (e.g. adenoviruses and enteroviruses), while others to a great extent remain bound to the cell surface (e.g. some herpesviruses), or even remain entirely intracellular (e.g. certain retroviruses and coronaviruses). Intracellular and cell-bound virus can be released by destroying the cell's integrity, using either cycles of repetitive freezing and thawing, or by ultrasonic treatment of the cells. Both techniques, however, may damage labile viruses. Virus enrichment and purification (*Table* 2) basically consists of two steps:

- removal of cellular detritus
- particle concentration

3.1 Centrifugation

As a first step in virus preparation, large cell culture constituents, that is, cells and cellular remnants, are removed by one or two low-speed centrifugation steps (4000–8000 *g* for 15 min), otherwise this debris will hide the structures of interest during TEM. As a second step, the virus contained in the resulting supernatant can be concentrated using one of several techniques, often applied sequentially.

3.2 Ultracentrifugation

Often a combination of low- and high-speed centrifugation, that is, differential centrifugation, is applied for purification and concentration. The virus from the cleared supernatants can be sedimented using a Beckman Airfuge™ directly onto the grid (100 000 *g*, 15 min). This technique is time-effective and diagnostically relevant, because it is appropriate for small volumes up to 100 μl. Larger volumes are processed using conventional ultracentrifuge sedimentation (100 000 *g*, 30–60 min). Labile viruses may be structurally protected by using sucrose cushion sedimentation. More effective, though also more time consuming, is enrichment by banding, and recovering the highly purified virus based on density (caesium chloride: 60 000–100 000 *g*, 24–48 h) or velocity gradients (sucrose: 60 000–100 000 *g*, 2–6 h). For detailed information see Mahy and Kangro (9) and Chapter 3.

3.3 Ultrafiltration

Ultrafiltration is used to concentrate and also to purify virus from suspension, by removal of small molecular detritus which can pass through the filter, for example, serum proteins. Filters are defined by their pore size, offering appropriate filter systems for different viruses. Depending on the complexity of the filter system, ultrafiltration can be very efficient in concentrating and purifying bulk volumes. As a drawback, the filter pores tend to become blocked during filtration, diminishing the filtration speed as well as the cut-off size. Likewise, molecular sieve chromatography removes small molecular detritus, but without concentrating the virus. Finally, agar filtration (*Protocol 2*) may be a helpful step before negative staining of small volumes. Agar filtration reduces both the volume and salt concentration, the latter often interfering with high quality staining.

3.4 Precipitation

Precipitation of viruses by salting out or solvent extraction may be used to reduce the volume of virus suspensions. By changing the properties of the solvent, its ability to suspend virus particles is lowered, and the virus falls out of suspension. Because many of the contaminating constituents are coprecipitated together with the virus, the resulting precipitate usually requires further purification before EM preparation.

3.5 Bioaffinity techniques

The high affinity of antigen–antibody reactions can be utilized for purification and concentration of virus suspensions. Virus-specific antibodies or other biospecific ligands are bound to a gel phase, and used either in affinity columns or batch systems. For virus purification and concentration at a preparative scale, affinity techniques are often not effective, as the methods employed to dissociate the virus–antibody complexes are often detrimental to virus structure. Integrated into EM preparation, however, affinity techniques are often successfully applied, for example, immunoaggregation in suspension followed by centrifugation or solid phase immuno-EM on the grid (*Protocols 4 and 5*).

4. EM grids and support films

EM grids, in particular the support films bearing the specimen, should fulfil three elementary requirements:

- chemical inertness
- mechanical stability
- electron transparency

Grids of various metals are used to stabilize the transparent support film. The commonly used copper grids are stable during handling, and cheap compared to other materials. Both sides of the grid can be used for filming, although they look and behave slightly differently. For negative staining, grids with small holes are recommended (e.g. 400 mesh, i.e. 400 holes per inch). In contrast, thin sections are mounted on grids with larger holes (100 or 200 mesh) in order to view the much larger cell structures. The grids are covered by two thin layers, one of plastic (20–40 nm) and the other of carbon (10–20 nm). This sandwich has a low electron scattering potential, that is, it is highly translucent to the beam. The plastic film offers mechanical stability. Commonly used plastics are collodion, or the more rigid polyvinylformal (Formvar, Pioloform). The additional carbon layer, by its high thermal conductivity, stabilizes the plastic film, and diminishes thermal drift as well as tearing. The surface properties of the grids, mainly charge and charge distribution, determine the quality of negative staining by influencing specimen adsorption and stain distribution. To improve particle adsorption to the support film, several pre-treatment methods are used, for example, glow-discharge, poly L-lysine, or Alcian blue treatment (*Figure 7*). Otherwise, the addition of small proteins, such as bacitracin or serum albumin, to the specimen may help to achieve higher adsorption rates.

Protocol 1. Preparation of grids for TEM

Equipment and reagents

- Carbon evaporation apparatus (e.g. Edwards High Vacuum International, West Sussex, UK)
- Fine pointed forceps (normal or reverse action) for handling grids
- EM grids; storage for the ready-made EM grids
- Filter paper (rectangular, ca. 3 cm × 6 cm) for removing the filmed grids
- Glass microscope slides
- Razor blade or scalpel
- Burette or wide angle bottle with filming solution (e.g. 1% Formvar or Pioloform in chloroform) (Polysciences Europe GmbH or Sigma)
- Dish with double-distilled water
- Acetic acid
- Acetone

Method

1. To remove contaminants (e.g. grease and other manufacturing traces), wash the grids in acetic acid for 1–3 min, then thoroughly with double-distilled water, and finally twice for 1–3 min with acetone.

Protocol 1. *Continued*

2. Clean a microscope slide using a clean linen cloth.
3. Stand the slide in the burette and let the filming solution flow out slowly. Let it drain vertically until the chloroform has evaporated.[a]
4. Using the razor blade, clean the four edges of the filmed slide.
5. Place the slide with the scored side uppermost onto the surface of a dish containing double-distilled water. Breathe on the film and allow it to float on the water surface by sinking the slide[b] (*Figure 8a*).
6. Place the grids in groups of 10–20 on the plastic film.
7. Cover the film with an appropriately sized piece of filter paper. When the filter paper is wet, remove it, with the grids adhering, from the water surface (*Figure 8b*). Let the grids dry protected from dust.
8. To stabilize the film against the electron beam, apply a carbon layer by vacuum evaporation, following the manufacturer's instructions. Monitor the carbon film thickness using a piece of filter paper, bent at an angle of 20–40° to the dish.
9. For hydrophilization of the carbon, use either glow-discharge (very effective by changing the surface charges; follow the instructions for the apparatus) or UV (30 min), or treat the film with 1% (w/v) Alcian blue or 10 mM poly-L lysine in double-distilled water for 5 min, followed by intensive washing.

[a] The stability, i.e. the thickness, of the plastic film mainly depends on two factors: first the concentration of the filming solution, and second the time the filmed slide is left in the solvent atmosphere for thinning.
[b] If the film does not come off the glass surface, grease the slide with a finger before filming.
[c] The hydrophilia of the glow-discharge or UV-treated grids is lost after some days; furthermore, repeated glow-discharge will thin the film considerably!

For high-resolution negative staining EM of purified viruses, special support films are used, consisting of a thin carbon layer supported by a holed plastic film and resulting in considerably higher brightness, contrast, and resolution (5).

5. Negative staining

Visualization of unstained virus preparations by conventional TEM is not rewarding. As biological specimens composed of light elements, they lack the electron scattering power required for high contrast imaging. Therefore, in the past, several staining techniques (e.g. osmium smoke treatment, heavy-metal shadowing) were developed in parallel with the development of EM, based also on electron-dense heavy metal staining. The negative staining technique has several origins and was broadly introduced into virology in the late

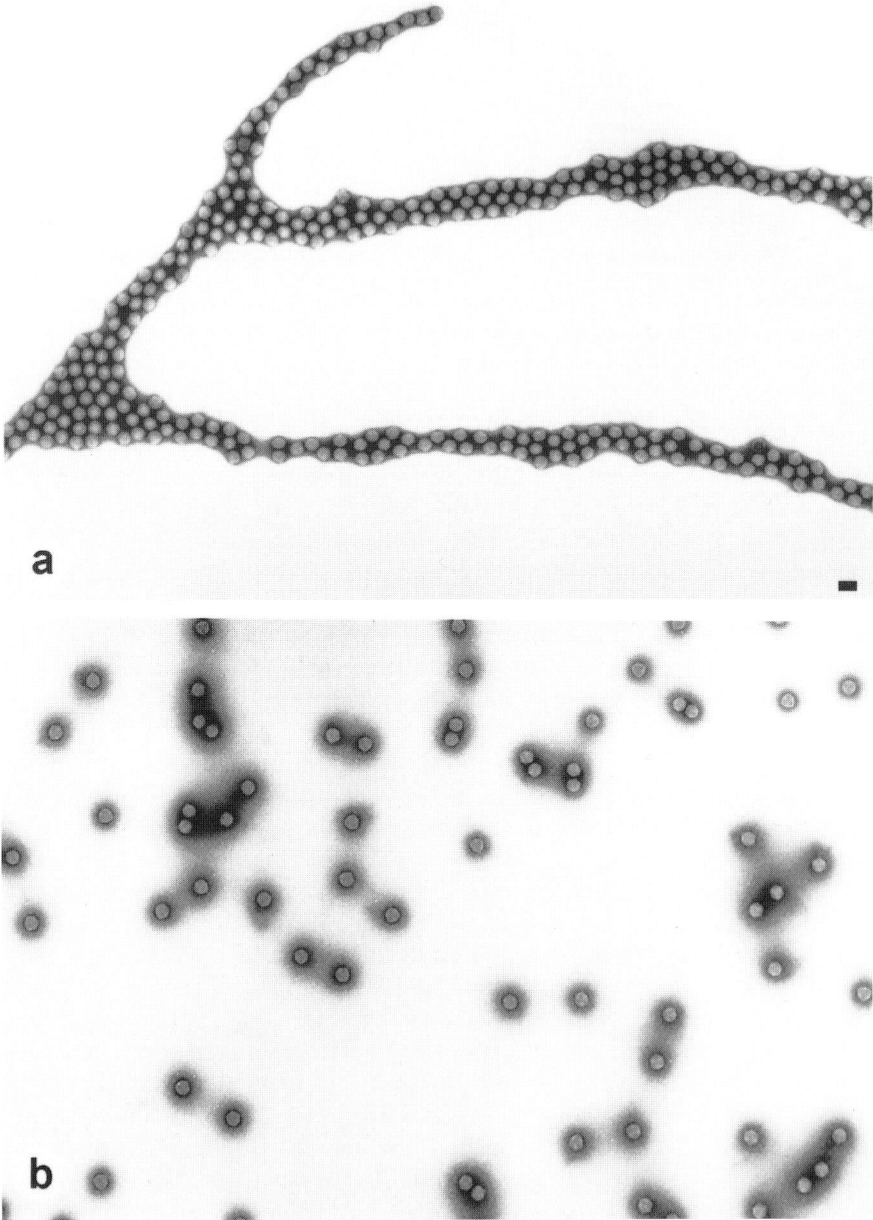

Figure 7. Influence of support film properties on the quality of negative staining. Depending on the charging of the grid, adenovirus and stain distribution differ significantly using (a) a 'native' grid without glow-discharge treatment, (b) after glow-discharge treatment. Two-step negative staining with 1% PTA (pH 7.0). Scale bar = 100 nm.

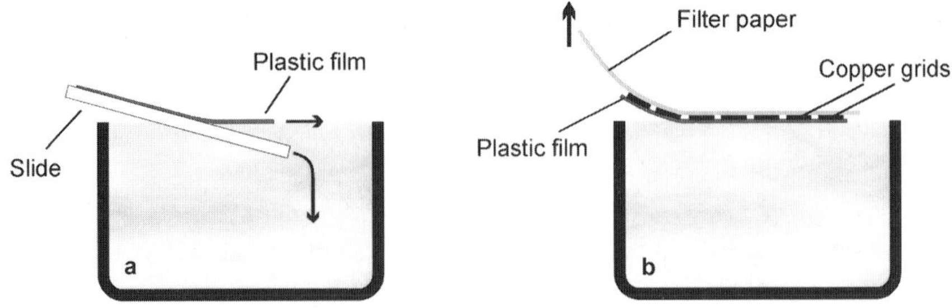

Figure 8. Preparation of plastic film-coated grids. (a) floating the film from the slide onto the water surface. (b) recovering the filmed grids from the water surface.

1950s (3, 10, 11). The term 'negative staining' was chosen as an opposite to positive staining applied in thin section EM. In negative staining, the stain does not bind chemically to the biological structure, for example, the virus, but merely surrounds it with a dried down, tightly fitting, amorphous glass-like layer of heavy metal salts. As a result, the particle appears light (i.e. electron translucent) in contrast to its periphery (*Figure 2*). Depending on the characteristics of the stain–virus system, in particular the stability of the virion, the particle reveals only outer structures (e.g. envelope, capsid, capsomeres), or the stain may penetrate into the particle, presenting inner structures too, for example, core, capsid layers, or nucleoprotein (*Figure 9*).

The staining and washing procedures used, in particular pH and osmolarity, can alter the structure of the virion. For example, the enveloped herpesviruses, influenza viruses, and retroviruses are more labile than the non-enveloped adenoviruses, papovaviruses, and parvoviruses, and as a rule of thumb smaller viruses are more stable than larger ones. Likewise, purified viruses are more labile than viruses contained in 'raw' cell-culture supernatants. Virus stability can be studied on grids by treatment with double-distilled water, detergents, or other destabilizing agents for varying time periods. This is also a convenient way to liberate the internally hidden RNP or capsid structures from enveloped virions.

In addition to structural alteration, further artefacts are associated with negative staining. In particular, with the two-step method, flattening of virus particles may occur, resulting in increased diameters, as well as a preferential adsorption of virus particles in defined orientations. Both effects are due to strong interactions between the virus and the charged grid surface, and can be diminished by several means, for example, applying only a weak glow discharge, or adding carbohydrate to the stain (e.g. glucose, trehalose) (10).

5.1 Negative stains: properties

Negative staining solutions are water-soluble, and after drying form a thin electron-dense amorphous layer without apparent crystallization. Most im-

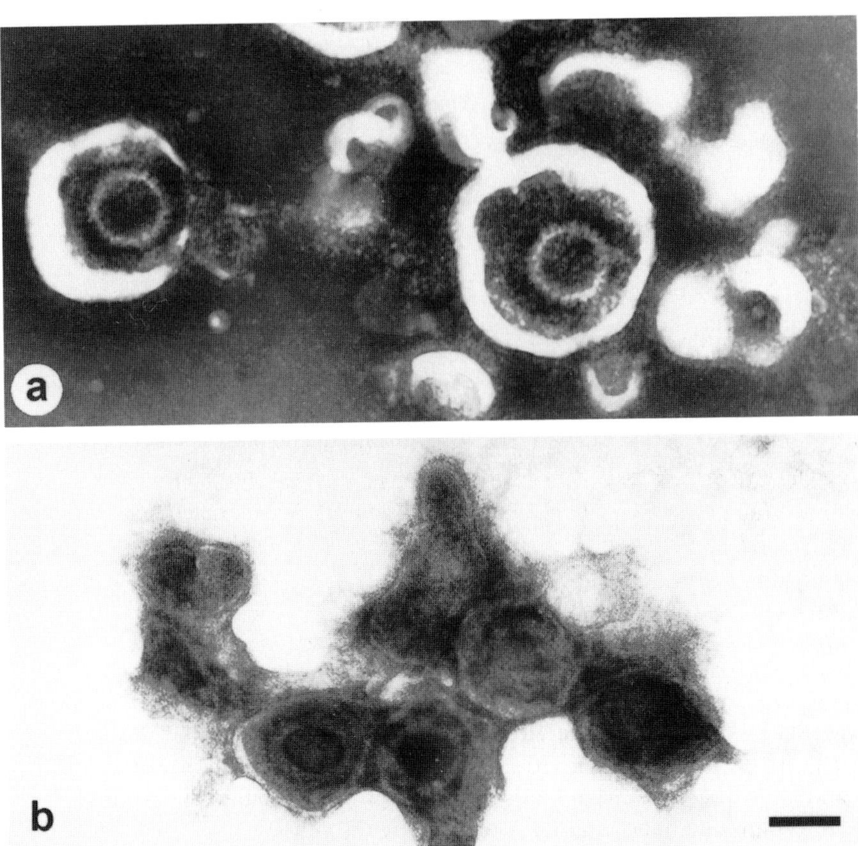

Figure 9. Aspects of herpesvirus using different negative stains. (a) After two-step negative staining with 1% PTA, pH 7.0, cell culture-derived enveloped viruses show distortion, as well as shrinkage of the envelope, surface projections and also the interior of the capsomeres, because PTA has penetrated the virus structure. (b) Herpesvirus negatively stained with 1% UAc, showing less inner structure due to the membrane-stabilizing properties of the stain. Scale bar = 100 nm.

portantly, they do not react extensively with virus structures under the conditions used. There are, however, distinct effects on virus membranes. While uranyl acetate (UAc), by its inherent positive staining effects, often stabilizes virus membranes, phosphotungstic acid (PTA) tends to destabilize virus envelopes. PTA often renders the virus envelope permeable to the stain (in particular at pH values over 7.5), thus permitting the demonstration of internal virus structures. This effect may be welcome, for example, for the demonstration of the capsid of herpesviruses or the paramyxovirus RNP, in particular with freshly prepared virus suspensions. UAc, in contrast, often leaves internal structures undetectable. Commonly used negative stains and their properties

Table 3. Negative stains commonly used for virus detection

Stain (concentration, w/v)	pH range	Properties and comments
Uranyl acetate, UAc (1–2%)	pH 4.2–4.5	Forms precipitates above pH 5.0 and in presence of phosphate; membrane stabilizing; high resolution; natural radioactivity.
Uranyl formate, UF (1%)	pH 4.5–5.0	Forms precipitates over pH 5.0 and in presence of phosphate; best available resolution; photolabile; natural radioactivity.
Phosphotungstate, PTA (1–2%)	pH 5.0–8.0	Broad useful pH-range; membrane destabilizing; tolerates relatively high salt concentrations; generates structural changes by interaction with many biomolecules (lipoproteins, membranes, lipids).
Silicotungstate, STA (2–4%)	pH 5.0–8.0	Similar to PTA, but higher in resolution.
Ammonium molybdate (2–3%)	pH 5.3–8.0	Broad useful pH-range; resolution sensitive to traces of buffers or salts; many materials sensitive to molybdate treatment due to cationic interactions; may act as mild chaotropic agent.

are listed in *Table 3*, and the effects using different stains with the same virus are shown in *Figures 9 and 10*. When searching for an unknown virus, it is often rewarding to use two kinds of differently acting stains in parallel, UAc or UF on the one hand, and PTA or STA on the other.

5.2 Negative staining procedures

A number of slightly different negative staining procedures have been described, with all of them based on two principal techniques: one-step and two-step procedures. Using one-step negative staining, defined volumes of the specimen and the stain are mixed for a defined time and placed onto an EM grid. After removal of excess liquid and drying, the grid can be evaluated directly. Though fast and simple to perform, this technique is not suitable for all kinds of specimens. Viscosity, purity, osmolarity, and charging properties of the specimen and the contaminants can prevent a successful TEM evaluation. The two-step procedure, including washing steps, is more versatile, for example, for the TEM analysis of (non-dialysed) gradient fractions, and non-purified, cell culture-derived, and diagnostic specimens. Interfering salt concentrations can be reduced by using different numbers of washing steps, different washing solutions, or agar filtration. Finally, agar filtration concentration and/or immunoaggregation or labelling can be included without difficulty.

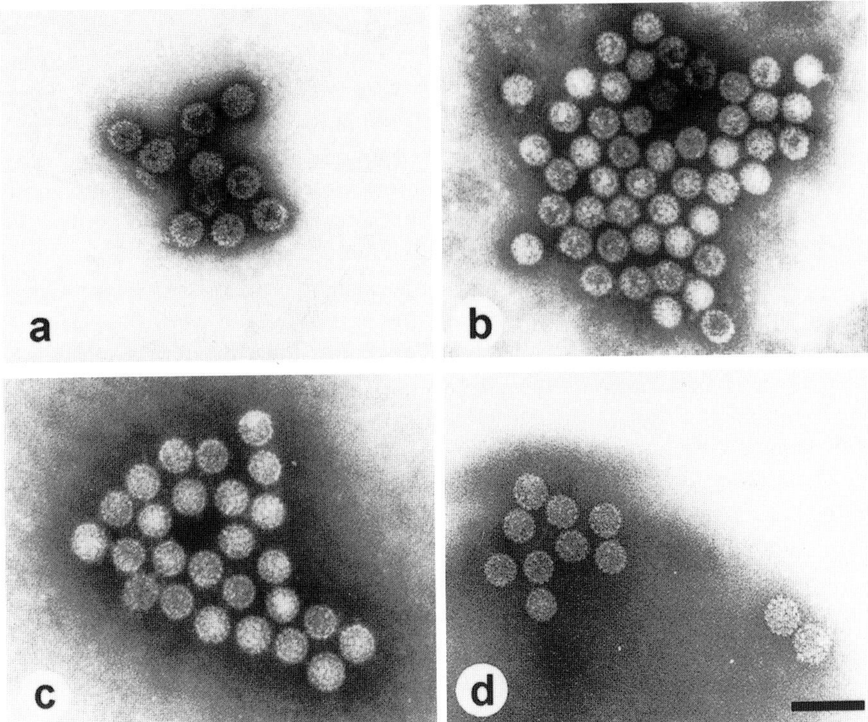

Figure 10. Comparison of different negative stains with the hamster polyomavirus. After two-step negative staining using (a) 2% UAc, (b) a mixture of UAc and PTA, each 0.5%, (c) 2% STA, pH 6.5, and (d) 1% ammonium molybdate, pH 6.0, subtle differences in resolution and contrast can be seen with this naked virus. Scale bar = 100 nm.

Protocol 2. Two-step negative staining

Equipment and reagents

- Prepared EM grids (*Protocol 1*)
- Parafilm™
- Fine pointed forceps (normal or reverse action) for grid handling
- Dish laid out with filter paper or equivalent storage for the stained EM grids
- Filter paper wedges, length 2–4 cm
- Washing reagent, e.g. double-distilled water or 0.05 M Hepes buffer pH 7.2
- Negative staining solution, e.g. 2% (w/v) uranyl acetate in double-distilled water[a]

1. Attach an appropriate length of Parafilm™ to the bench using a little water as 'adhesive'.
2. Place the reagents as drops (at least 10 μl, better 30–50 μl) in line onto the Parafilm™: one drop of sample solution, several washing drops,

Protocol 2. *Continued*

and one drop of stain (the number of washing drops varies from 2–5 depending on purity and salt/sucrose concentration of the sample).

3. Float one or several EM grids with the filmed side on the sample for 1 min (adsorption time may vary from 3 s to 10 min depending on material concentration. Extra grids may be stored for convenience on the sample drop, and processed after checking the first grid by direct EM examination (allows a quick orientation).
4. Wash the grid by floating onto the first washing drop for 2–10 s, and drain the fluid by carefully dabbing the grid vertically onto a dry filter paper wedge. The specimen should never dry before the procedure is finished.
5. Repeat the steps with each washing drop. Washing will inevitably reduce the number of particles on the grid. Therefore a compromise has to be found between particle content, purity, and stain distribution on the grid.
6. For staining, float the grid on the stain for 3–60 s, depending on stain and specimen; remove the fluid by dabbing the grid on to the damp edge of a filter paper wedge, and allow the grid to air-dry. The stain thickness depends on the dabbing angle due to capillary forces. There is also stain between the prongs of the forceps. This fluid should be carefully removed using the tip of a filter paper wedge to avoid reflux of the stain on to the grid.

Note: To save material and time while checking the quality of the preparation in the EM, cover the row of drops with a lid and add some wet tissue to prevent the drying of the drops (humid chamber).

[a] Avoid phosphate ions and or incident light during incubation of specimens with UAc.
[b] Specimen adsorption—alternatives to 3:

I. *Direct examination* Remove the culture medium from the culture, apart from a thin liquid film, and slightly press the grid direct onto the cell layer.

II. *Agar filtration* Pipette a drop of virus suspension (50–80 μl) onto the surface of a 1.5% bacteriological nutrient agar plate, and cover it with an EM grid. When the liquid is almost absorbed totally by the agar, remove the grid.

5.3 Quantitative aspects: measuring and counting in TEM

In negative staining, there is always interaction between the virus and the grid surface, which influences virus structures, for example, by flattening or shrinking, as well as osmotic disruption of particles. Pre-fixation or on-grid fixation of the specimen by 0.1% (v/v) glutaraldehyde may provide better preservation, in particular with purified virus particles. Also, the TEM itself may show instrumental aberrations in magnification. Therefore, EM-based quantitative data should be collected and assessed with care. An internally controlled solution for determining particle sizes is given by the addition of an internal marker with known dimensions, for example, catalase crystals (lattice spacing 8.75 nm, *Figure 11*). By measuring both the virus and the marker (on the same EM negative using a ×10 measuring magnifier) and bringing the marker values

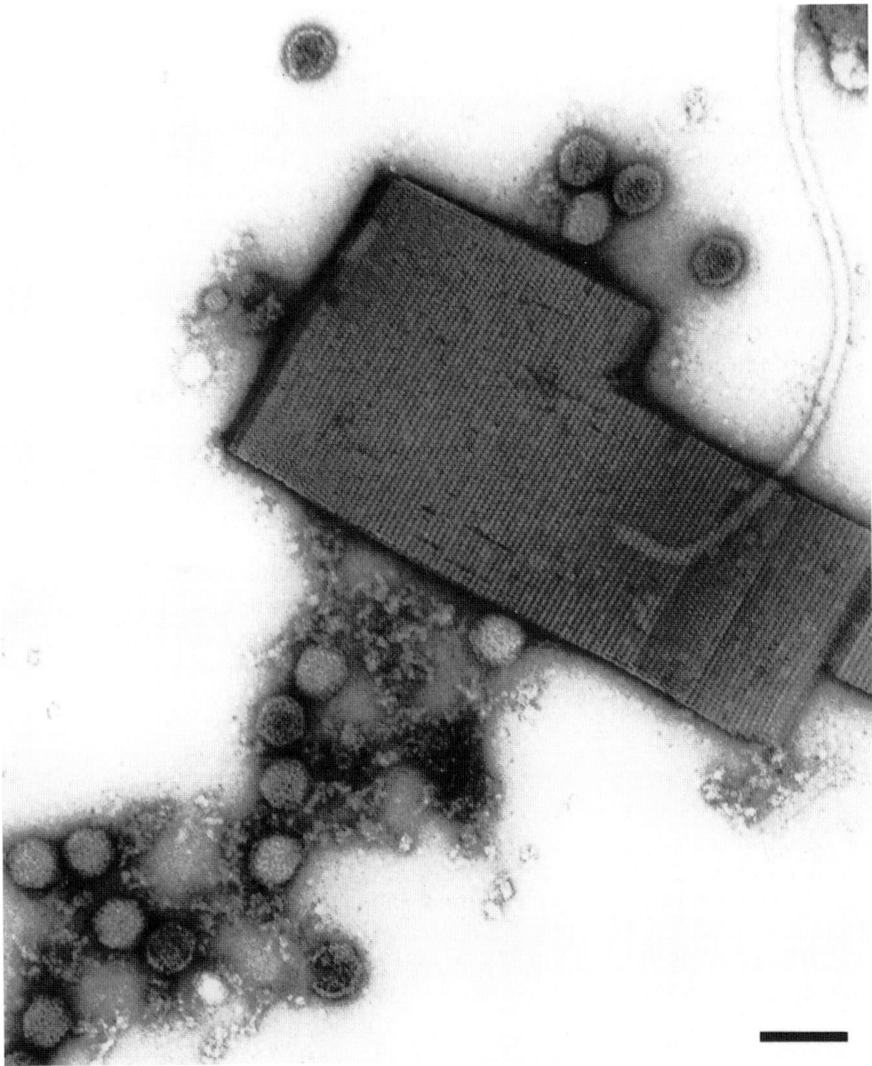

Figure 11. The use of catalase crystals as an internal length standard. From the marker's known lattice spacing of 8.75 nm, the diameter of the eel birnavirus was determined to be 61–65 nm. The specimen mixture was two-step negatively contrasted with 1% UAc. Scale bar = 100 nm.

in line with the known standard values, virus dimensions can be determined with a deviation of less than 5% (10, 11).

The same principle, use of an internal marker, is also applicable to the problem of particle counting. After mixing defined volumes of the virus suspension and a suspension of latex beads with known concentration, the respective particles can be counted after negative staining or thin section TEM. The concentration of the virus is then assessed by comparison with the latex bead concentration.

6. Embedding and thin sectioning

In conventional TEM, electrons are used at accelerating voltages of 60–150 kV. Because of their limited penetration power, only a few biological objects, such as viruses and cellular organelles, are thin enough to be penetrated and visualized adequately by these electrons. To analyse the structure of thicker objects, for example, virus-infected cells, two methods are used:

- use of electrons accelerated to mega-electron voltages (MeV)
- use of ultrathin sections ranging in thickness between 30–80 nm (*Figure 12*)

The MeV-EM required for the first method is both inconvenient to use and unavailable to most laboratories. In addition, the image generated from thick specimens is usually very complex and, because of the high information content, it often requires significant image processing efforts.

Ultrathin sectioning, on the other hand, has been developed into a well-established and reliable tool in TEM. The section thickness of 30–80 nm is comparable to the size of many viruses. Therefore, virus particles contained in a thin section can be presented with relatively high contrast and resolution. The preparation for thin section TEM consists of four essential procedures:

- fixation
- embedding
- sectioning
- staining

These steps are in part intertwined, and all of them are crucial for the preservation of fine structure, resolution, and appreciable contrast (*Protocol 3*). Note that what is seen in positive contrast EM is not the specimen itself, but the differentiated distribution of different positive stains, for example, heavy metals.

6.1 Fixation and staining

The goal of fixation is to preserve as much as possible the structure of living matter during preparation and evaluation of the ultrathin section. Because of the high attainable resolution, fixation for TEM is definitely more demanding

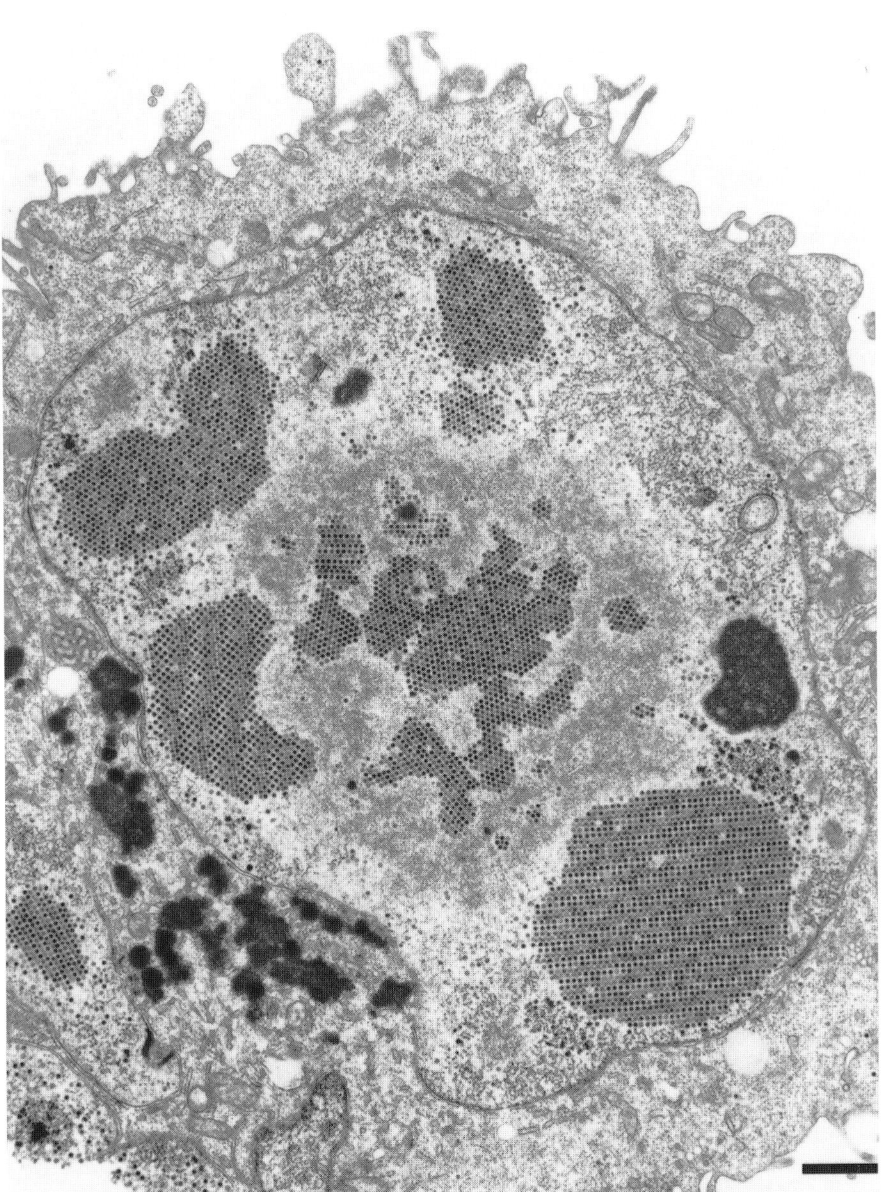

Figure 12. Structure of an adenovirus-infected cell. Thin section showing the assembly of adenovirus in the nucleus of an infected HEL cell. The virus particles are densely packed, forming three-dimensional crystals with peculiar patterns, because the section plane cuts the crystals at different orientation angles. The section thickness (about 50 nm) is less than the diameter of the virion (80 nm). Scale bar = 1000 nm.

than for light microscopy. At present, rapid freezing techniques are optimum for this task, although only for very small specimens (2, 3, 4, 10, 12). Furthermore, this type of physical fixation is usually followed by rather complex procedures, such as cryo-sectioning and cryo-EM, or freeze-substitution followed by low temperature resin embedding.

Chemical fixation, on the other hand, consisting of a routine sequential double-fixation scheme with glutaraldehyde (GA) and osmium tetroxide (OsO_4), is easy to perform (3, 4). While GA, as a di-aldehyde, cross-links the amino-groups of proteins in particular, OsO_4 mainly binds to unsaturated fatty acids. In contrast to GA, OsO_4, because of its high atomic weight, also acts as a stain. Double fixation can be performed also as a one-step procedure that, however, may be prone to failure. Chemical fixation is performed by mixing the specimen, that is, a suspension of cells with the fixative, at 37 °C (or at least at room temperature). With monolayer cells, the fixative can be mixed directly into the culture medium. Physiological temperatures, small specimen size, and in critical cases the application of GA-formaldehyde mixtures help for a rapid penetration of the fixative which definitely is a prerequisite for a well preserved fine structure. The pH and osmolarity of the buffered fixative are further determinants: only after post-fixation by OsO_4 does the specimen become insensitive to osmotic changes. As a further incubation step, 1–2% uranyl acetate helps to produce a clear, high-contrast presentation of membrane structures (see also *Figure 3*). Finally, to improve further the overall contrast, an incubation with tannic acid renders many cellular components (in particular the difficult-to-stain carbohydrates) susceptible to an intense lead post-staining. Additional information on commonly used fixatives and their properties is listed in *Table 4*.

6.2 Embedding

If not processed *in situ*, adhering to a small piece of the cell culture plate (e.g. Terasaki microtiter plate (Nunc) or Thermanox™ plate (Heraeus Instruments)), after OsO_4 treatment cells are enclosed in agar for ease of handling, and processed further during dehydration, infiltration, and embedding as agar blocks. Embedding schemes can be varied according to the specific requirements of the specimen, for example, proper embedding of bacteria and plant tissue requires prolonged dehydration and infiltration periods, and the choice of a low viscosity resin (because the cell wall represents a relative diffusion barrier). There are several mixtures of epoxy resins in use for embedding (e.g. Araldite, Epon 812 or respective substitutes, Spurr's low viscosity epoxy resin (Polysciences Europe GmbH or Sigma)) differing in viscosity and the hardness of the final resin block (3). Epon-embedded specimens shrink less than 5%, which is much less than with all other resins. The hardness can be controlled over a wide range by changing the ratios of the resin constituents. To allow identification of the specimens, small labels are placed together with the blocks into the embedding capsule, which is finally filled up with the resin

Table 4. Commonly used fixatives for embedding and thin sectioning

Fixative	Fixation properties/ fixed structures	Antigen preservation	Positive staining
Physical fixation			
Water (amorphous ice)	Fixation is limited by water crystallization (depth $\leq$ 200 μm); fixes all structures.	++	–
Chemical fixation			
Glutaraldehyde (GA)	Cross-linker, two reactive aldehyde groups; mainly fixation of structural and soluble proteins.	– –	–
Formaldehyde (FA)	Monoaldehyde, cross-links by forming active polymerization products; mainly fixation of structural and soluble proteins, fixation less stable compared to GA.	+	–
Osmium tetroxide	Cross-linker, reactive group: tetroxide; mainly fixation and stabilization of lipid membranes in co-operation with GA and/or FA.	–	+
Uranyl acetate (UAc)	Cross-linker, reactive group: dioxide; fixation and stabilization of membranes and nucleic acids.	–	+

mixture. Epoxy resins are cured by heat (less commonly by UV light), leading to extensive cross-linking and hardening of the resin components. Many of the preparation steps – dehydration and infiltration in particular – can be performed automatically, using specially designed rotating tissue processors, for example, Lynx (Leica Microsystems) (2, 3, 4, 12).

6.3 Thin sectioning

For cutting thin sections, well cured specimens are required (compare the consistency of the resin block to the hardness of a thumbnail). Ultrathin sections – in the range from 20 to 40 nm – require relatively hard Epon blocks, which can be generated by the selection of appropriate resin mixtures. Using today's equipment, ultramicrotomes and diamond knives, the sectioning process is no longer experimental work. Glass knives, if carefully prepared, may be used, too. Before starting to cut ultrathin sections, the tip of the specimen block containing the specimen is trimmed to a flat pyramidal surface of a proper size and shape. Trimming can be performed manually using razor blades, or semi-automatically using trimming machinery and, for fine trimming, also the ultramicrotome. The use of survey sections of 0.5–2 μm in thickness and up to 2 mm in edge length allows an effective light microscopical control of the embedded specimen. Semithin sections, after staining, are perfectly well suited for high resolution light microscopy, and thus represent a high quality tool in correlative microscopy.

After assessing orientation and trimming down to a smaller size, ultrathin sections of the interesting part of the specimen are prepared and collected on 'naked' grids (without support film, to achieve higher contrast) or, for more stability, on carbon-reinforced plastic-coated 100 or 200 mesh grids. The technical parameters and methods in thin section EM (such as varying knife angle and clearance, the handling of diamond knives, staining methods for semi- and ultrathin sections) are best learned during practice in a laboratory staffed with experienced personnel.

6.4 Post-staining

In spite of two positive-staining steps (OsO_4 and UAc) during embedding, ultrathin sections must be post-treated in order to generate appreciable contrast. Post-staining is achieved by treating the sections on grid with uranyl acetate (if not part of the embedding schedule), followed by routine lead citrate staining. Special care must be taken to avoid CO_2 contamination during UAc and lead staining (3, 4, 12). The contrast behaviour also depends on section thickness: although the resolution attained in very thin sections often is very welcome, it may be difficult to reach a balanced contrast because only a little material is contained in the thin section. Generally, the use of a lower acceleration voltage, for example, 60 keV instead of 80 keV (as suitable for negative staining) increases contrast, and is often beneficial in the evaluation of thin sections.

Protocol 3. Cell embedding, sectioning and post-staining

Equipment and reagents

- Buffer: 0.1 M PBS or 0.05 M Hepes, pH 7.2
- Fixative: glutaraldehyde, 2.5% and 5% (v/v) in buffer (at room temperature or 37 °C)
- Osmium tetroxide OsO_4, 1% (w/v) in double-distilled water
- Low melting point agarose, 3% (w/v) in double-distilled water, adjusted to 37°C (water bath)
- Uranyl acetate, 2% (w/v) in double-distilled water (degas the water by boiling)
- Tannic acid, 0.1% (w/v) in buffer, sodium sulfate Na_2SO_4, 1% (w/v) in buffer
- NaOH, 0.02 N for washing in post-staining
- Lead citrate, 0.4% (dissolve 0.1 g lead citrate in 25 ml boiled double-distilled water; afterwards add 0.25 ml 10 N NaOH to avoid precipitation of lead carbonate)
- Different ethanol concentrations: 30%, 50%, 70%, 95%, 100% (v/v) in double-distilled water
- Resin (e.g. Embed 812 (Polysciences Europe GmbH, or Electron Microscopy Sciences, Fort Washington, Pa, USA)); propylene oxide or styrene as solvent; resin–solvent mixtures 1: 1 and 1: 2
- Filter paper, scalpel, fine-pointed forceps, cell scraper
- Rotation unit to turn the vials during dehydration and infiltration
- Gelatin or polyethylene capsules, or flat embedding moulds (specimen containers)
- Water bath (adjusted to 37°C), heater for resin polymerization

Method

A. *Fixation, dehydration, and embedding*

1. Monolayer cells: remove the culture medium, leaving a thin liquid film on the cells. Add 2.5% glutaraldehyde directly into the culture

dish, and after 30 min remove the cells from the bottom using a cell scraper.

Suspension cells: collect the cells by centrifugation and mix equal volumes of the cell suspension and 5% glutaraldehyde; fix for 30 min.

2. Gently centrifuge the cells (500 *g*, 10 min), discard the supernatant, and resuspend and re-fix the cells with 2.5% glutaraldehyde for 1 h. Before embedding, glutaraldehyde-fixed cells may be stored in the fixative for several weeks at 4°C without appreciable loss in fine structure.
3. After glutaraldehyde fixation, sediment the cells and wash the fixative out by gentle centrifugation, resuspend the cells for 5 min in buffer and sediment again, and discard the supernatant.
4. Post-fix the cells by suspending in a small volume (1–2 ml) of OsO_4 for 1 h in the dark. Sediment the cells and discard the supernatant. Afterwards, wash the pellet twice with double-distilled water.
5. Add an equal volume of the melted agarose to the cell sediment (in the water bath), gently mix and cool down the tube. Cut the agarose clot, after setting, into cubes of 0.5–1 mm edge length.
6. Treat the agarose blocks with tannic acid for 1 h; wash twice for 5 min with Na_2SO_4, followed by two washes with double-distilled water (10 min each). To increase penetration and exchange of reagents and solvents, perform steps 6–9 using a rotation mixer.
7. Treat the agarose blocks with UAc for 1 h in the dark; afterwards wash twice with double-distilled water (10 min each). Avoid phosphate ions and incident light during incubation of specimens with UAc!
8. Dehydrate the blocks in a graded ethanol series as follows: 30%, 50%, 70% each for 15 min; 95% for 20 min; 100% twice for 30 min.
9. After dehydration, infiltrate with resin-solvent mixture (1:2) for 2 h, then with resin-solvent mixture (1:1) overnight, and finally infiltrate with pure resin for 3 h.
10. Place the agarose blocks in capsules with fresh resin and polymerize for 2 days at 60°C. For rapid processing protocols, e.g. using Spurr's resin, see Bozzola and Russel (3) or Robards and Wilson (4).

B. *Sectioning*

The complexity of the parameters for thin sectioning depend on the ultramicrotome itself, the resin, and the properties of the specimen, the knife, and the trough to collect the sections. Therefore it is not possible to give a thorough description of the procedure here. Briefly, the cured resin block is fixed in special specimen holders, and trimmed to assess the specimen.

Protocol 3. *Continued*

After light-microscopic control of semithin sections and fine trimming, the specimen is cut with diamond, sapphire, or glass knifes to give sections of 30–80 nm. The sections may be flattened using heat or solvent vapour, and kept on naked or, for post-embedding labelling, filmed copper grids. Before TEM evaluation, the 'naked' sections are reinforced by a thin layer of carbon.

C. *Post-staining of ultrathin sections*

1. Place the grid for 0.5–2 min into a drop of lead citrate (0.4%).
2. Wash immediately by briefly immersing the grid into three drops of 0.02 M NaOH.
3. Afterwards thoroughly rinse with double-distilled water and quickly blot the grid dry between two sheets of filter paper.

7. Immuno-electron microscopy (IEM)

To combine the analytical potential of antigen–antibody reactions with the high resolution of TEM, several immuno-EM (IEM) techniques have been developed (*Table 5*). IEM is applied with the aim of identifying the antigen or

Table 5. Comparison of immuno-EM (IEM) techniques

Method	**Field of application**	**Properties**
Negative staining IEM	Antibody evaluation, antigen detection	Fast, easy to use
Disperse immuno-EM (*on grid*)	Principal detection of antigens: Evaluation method for antibody-antigen systems	Labelling not always distinct
Immuno-aggregation (*in suspension*)	Enrichment of viruses, first evaluation method for antibody-antigen systems	Most reliable, simple, and fast
Immuno-decoration	Localization of antigens	Difficult to handle on grid
Solid-phase immuno-EM	Enrichment of viruses, virus diagnosis	Difficult to establish; on grid, liable to break down
Thin-section IEM	Detection and localization of plastic sections	Cellular and virus antigens
Pre-embedding labelling	Detection of external antigenic determinants	Sensitive, but limited
Post-embedding labelling	Detection of internal and external antigens	Most versatile
Labelling of cryo-sections	Detection of internal and external antigens	Very sensitive, though more difficult to perform than post-sectioning labelling

antibody, or for immunospecific virus enrichment, or for the localization of virus or cellular antigens. The characterization of antibody specificity and immuno-enrichment are often applied together with negative staining. Localization of cellular antigen is done using thin sections.

In IEM, two principal labelling techniques are differentiated:

- direct (one-step)
- indirect (two-step)

Both techniques have in common the first step: incubation of the specimen with a specific antibody (derived from a monoclonal cell culture, or from a hyperimmune serum) which binds to the antigen of interest. Using direct labelling, the bound antibody is recognized as a smooth layer surrounding the virion, or by an electron-dense marker (e.g. colloidal gold), provided the primary antibody is tagged. For indirect labelling, a second antibody or protein A, specifically binding to the Fc fragment of the first antibody and coupled to an electron-dense marker, is incubated with the specimen to bind immunospecifically. Using indirect labelling, a single anti-IgG antibody reagent directed, for example, against mouse IgG, can tag all primary antibodies derived from mouse or mouse hybridoma cells. Different antigens can be detected simultaneously in one specimen by using different markers, for example, colloidal gold with two significantly different diameters.

7.1 Antibodies: properties and usage

In IEM, three main types of antibodies or functionally related molecules are in use. Hyperimmune sera are raised in animals, for example, mouse, rabbit, or rat, immunised with the specific antigen (Chapter 7). Sera contain specific as well as non-specific antibodies from all possible immune globulin subclasses, and also other serum proteins and other compounds, for example, lipids. Serum antibodies are polyclonal, detecting a variety of different antigenic domains. Early in the immune response, the pentavalent IgM antibodies are raised, but later on only the bivalent IgG is detectable. Monoclonal antibodies are cell culture-derived, monospecific for a single determinant, and belong to only one immunoglobulin subclass, mostly of the IgG-family (Chapter 7). The properties of monoclonal antibodies can be precisely described, but their affinity is often up to 10^2 times lower than with corresponding hyperimmune sera. Protein A is a cell wall protein from *Staphylococcus aureus*, which specifically binds to the Fc portion of IgG antibodies from almost all mammalian species. Although there are five IgG-binding sites present on protein A, in practice a maximum of three IgG molecules can be bound (13).

For storage, the antibody stocks should be mixed with glycerol at an antibody-glycerol ratio of 1:2, divided into adequate portions, and frozen at −20 °C. At this temperature the glycerol will prevent freezing, while the low temperature prevents any biologically detrimental processes. For use, the

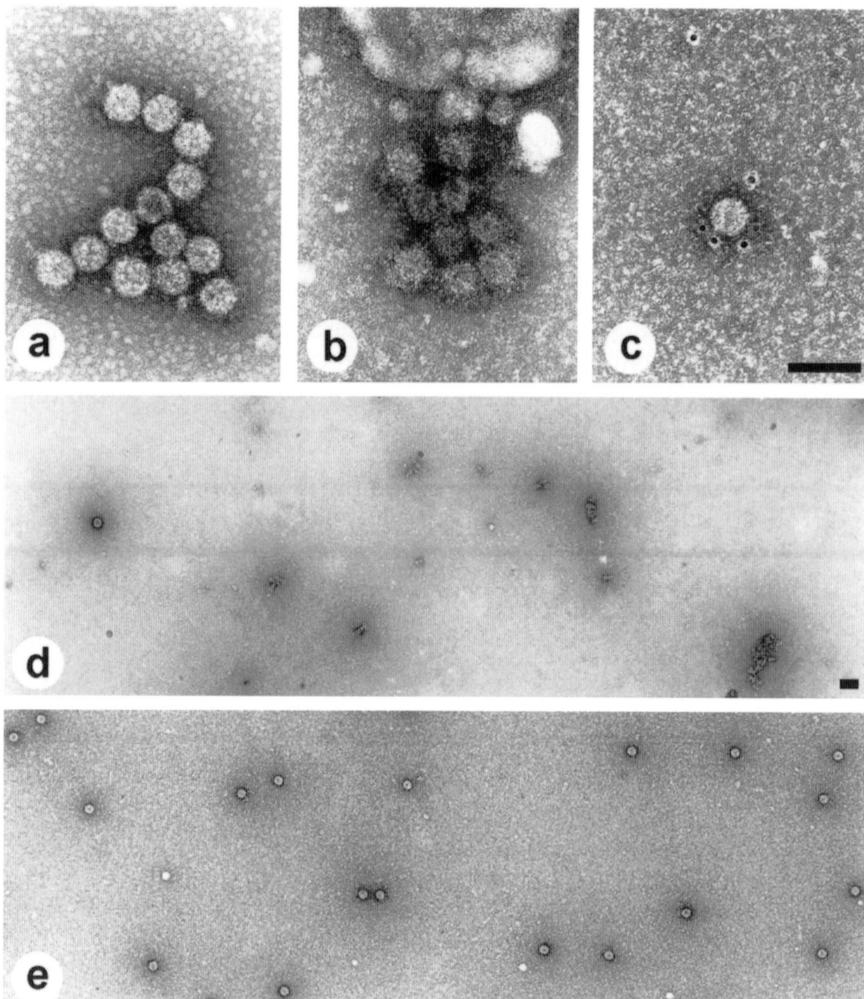

Figure 13. Different immuno-negative staining techniques applied with polyomavirus BK. (a) The immuno-aggregate shows defined distances between the single particles (b) Disperse immuno-EM: the viruses incubated on-grid are surrounded by a corona of antibodies. (c) For demonstration of the primary antibody binding, gold-labelled anti-IgG antibodies are bound to the primary antibody surrounding the virus. (d), (e) Solid phase immuno-EM: while in the negative control (d) the normal, untreated urine shows only one virion (upper left), in (e), the SPIEM specimen, virus particles are concentrated and, in addition, contamination is significantly decreased. Scale bars = 100 nm (two-step negative staining using 2% UAc).

antibody aliquots should be diluted to the final concentration using a buffer containing small molecular weight protein, for example, PBG (*Protocol 4*). Besides the structural and functional protection of the antibody, PBG will also avoid non-specific antibody reactions on the grid ('conditioning' of the grid). When working with a new (i.e. not previously used for IEM) antibody, a suitable antibody concentration has to be pre-determined. For the initial screening, three dilutions of the antibody (prepared in PBG) should be tested, for example, 1: 10, 1: 100, and 1: 1000. An easy and reliable test system for antibody evaluation is immuno-aggregation (*Protocol 4B*). Most important in IEM, however, is the preparation of several controls together with the specimen, that is, antibodies unrelated to the antigen and negative specimens lacking the antigen. To achieve reliable information on the specificity and background labelling, it often helps to use and to analyse two or three different antibody concentrations in parallel.

7.2 Immuno-negative staining

The field of immuno-negative staining applications is wide ranging, from diagnostics to basic research problems. This technique is used to visualize viruses better from diluted and/or contaminated suspensions, by formation of readily detectable immuno-aggregates, and for mapping virus surface determinants using specific antibodies, for example, disperse immuno-EM, immuno-decoration (*Figure 13*). Basic immuno-negative staining techniques are described in *Protocol 4*. Immuno-aggregation involves the cross-linking of viruses by mixing the specimen with bivalent IgG antibody. These aggregates are more easily detectable due to their size, and can be further concentrated by centrifugation. To distinguish immuno-aggregates from 'normal' aggregates, it helps to judge the distances between the particles: in immuno-aggregates the particles are separated by the distance given by the antibody binding sites. Immunolabelling is the direct labelling of virus on grid, and can be performed as a direct or indirect labelling procedure. Because the virus is first adsorbed to the grid, aggregation cannot occur. Instead of cross-linking viruses, the antibodies build a smooth layer around them. This layer can be detected after negative staining. In high resolution cryo-EM, single specific antibodies or Fc fragments are used as a tool in basic research.

Protocol 4. Immuno-negative staining (basic techniques)

Equipment and reagents

- Negative staining equipment and reagents (*Protocol 2*)
- 0.1 M PBS, pH 7.2, containing 0.2% (w/v) gelatin (PBG)
- Specific antibody and anti-species IgG, or protein A-gold conjugate (diluted in PBG)
- Humid chamber, kept at 37°C

Protocol 4. *Continued*

A. *Disperse immuno-labelling EM (on grid)*

Apply the following procedure between steps 5 and 6 of *Protocol 2*. The grid must not dry after the first antigen or antibody steps, otherwise non-specific adsorption will occur, obscuring the antigen–antibody reaction.

1. To condition the grid, float it on two drops of PBG for 5 min each.
2. Float the grid on the specific antibody for 20 min to react with the pre-adsorbed virus.
3. Remove excess antibody by washing on three drops of PBG.

B. *Immuno-aggregation (in suspension)*

1. Mix defined volumes of specimen and specific antibody in a plastic reaction tube (1.5 ml) and incubate the mixture for 1 h (alternatively overnight at 4 °C).
2. For concentration of immuno-aggregates, centrifuge the suspension for 10 min at 10 000 *g*, and resuspend the pellet in a small volume of PBG.
3. Perform negative staining according to *Protocol 2*.

C. *Two-step immuno-gold-labelling (on grid)*

Perform the following steps after step 4 of *Protocol 2* (negative staining). The grid must not dry after the first antigen or antibody steps, otherwise non-specific adsorption will occur, obscuring the antigen–antibody reaction:

1. To condition the grid, float it on two drops of PBG for 5 min each.
2. Float the grid on a drop of specific antibody for 30–60 min at 37 °C, and wash on two drops of PBG for 10 min each.
3. Float the grid on a drop of the second, labelled antibody for 30 min at 37 °C, and wash on two PBG drops for 5 min each.
4. Float the grid on one drop of PBS for 2 min, followed by 2 drops of distilled water for 1 min each.
5. Continue with step 6 of the negative staining procedure (*Protocol 2*).

Solid-phase immuno-EM (SPIEM) is a very specific though complex technique. It is based on grids coated with specific antibody, bound either directly by non-specific adsorption, or specifically, for example, by using protein A. While the antibody-coated grid is floated on the specimen, the virus may be immuno-specifically captured and bound to the grid. The advantages of SPIEM compared with normal negative staining are the specific enrichment of the antigen (20–1000-fold, depending on the virus-antibody system), and the decrease of contamination on the grid. Unpurified specimens with low virus concentrations, in particular, can be more easily assessed, which is especially helpful in virus diagnosis (*Protocol 5*). However, SPIEM can only be successfully performed if high affinity specific antibodies are available, and if the system parameters, for example, antibody dilution, adsorption times, feasible pH range, etc., are carefully determined (13).

Protocol 5. Solid phase immuno-electron microscopy (SPIEM)

Equipment and reagents

- Immuno-negative staining equipment and reagents (*Protocol 4*)
- Hotplate, adjusted to 35–40 °C
- Fresh protein A solution (5 μg ml^{-1} in PBG, *Protocol 3*)

Method

1. Attach an appropriate length of Parafilm™ to the bench using water as adhesive, and attach another piece of Parafilm™ to the hotplate.
2. Place the protein A and the antibody solution as drops (30–50 μl) on the bench Parafilm™, and the sample drop (50 μl) on the hotplate Parafilm™.
3. Float an EM grid on protein A for 15 min, and drain excess fluid using a filter paper wedge. No rinsing is required. The grid must not dry after the protein A, or following antibody incubation steps.
4. Float the wet grid on the antibody for 30 min, and remove excess fluid as described above.
5. Float the antibody-coated grid on the sample drop at 35 °C for 30–60 min.
6. Apply steps 4–6 of the negative staining procedure (*Protocol 2*).

7.3 Immuno-labelling of thin sections

As stated above, the well conserved fine structure of cells observed in thin section TEM is based on routine glutaraldehyde-OsO_4 double fixation, intense cross-linking of the resin, and several staining steps utilizing a number of heavy metals. This regime, unfortunately, interferes with the properties of living matter, for example, antigenicity and enzyme activities. In conventional ultrathin sections, glutaraldehyde cross-links and blocks amino groups present in many antigenic determinants, while OsO_4 destroys antigenicity, mainly by oxidation. Also, the high degree of cross-linking of the epoxy resin prevents the antibody from binding the corresponding antigenic determinants. Therefore, to study enzymes or antigens (or respective antibodies) by thin section TEM, suitable fixation and embedding techniques have to be employed.

The conditions should be chosen in a controlled way, to approach a tolerable compromise between fine structure and preservation of biological activity. Usually, immunofluorescence on semi-thin sections of the embedded specimen is sufficient to assess the preservation of antigens. As a rule of thumb, the sensitivity of immunofluorescence parallels that of IEM. Therefore checking the system by the less demanding immunofluorescence is strongly recommended.

To preserve antigen reactivity, the amount of glutaraldehyde, as well as the fixation time, must usually be reduced. Depending on the antigen, different fixation conditions have been worked out experimentally. In order to keep a decent fine structure in IEM, and to avoid redistribution of antigens during immune incubations, low concentrations of glutaraldehyde are necessary to prefix the specimens in a controlled way. A mixture of 0.025% glutaraldehyde and 4% formaldehyde applied for 10 min at room temperature is often sufficient. To stop aldehyde activity, blocking buffers are used containing lysine, glycine, or NH_4Cl.

If specimens are embedded and processed in methacrylates or as ultrathin cryo-sections (instead of the highly cross-linked epoxides), antigens are more readily accessible. For post-sectioning immuno-labelling, the low viscosity methacrylates (Lowicryl, LR White, Unicryl) are particularly well suited. Being hydrophilic, they are cured for IEM preferentially at low temperature by UV. Finally, to study exclusively surface determinants on cells or viruses, pre-embedding IEM is often preferred. After a weak and brief glutaraldehyde pre-fixation (see above), immunolabelling is performed, followed by conventional resin embedding. As an alternative to methacrylate sections and post-sectioning IEM, the highly sensitive, though technically more demanding labelling of ultrathin frozen sections may be used (3).

Protocol 6. Immuno-labelling of thin sections

Equipment and reagents

- Buffer: 0.1 M PBS, pH 7.2, or 0.05 M Hepes, pH 7.2
- Fixative: prepare fresh formaldehyde from paraformaldehyde and dissolve 4% formaldehyde (v/v) and 0.025% glutaraldehyde (v/v) together in buffer
- Glutaraldehyde 2.5% (v/v) in buffer
- 50 mM glycine or lysine in PBG (to block free aldehyde groups)
- 0.2% (w/v) gelatin in PBS (PBG)
- Specific antibody and anti-species IgG, or protein A-gold conjugate (diluted in PBG)
- Terasaki microplate, treated with 1% (w/v) Alcian blue for 30 min to improve cell adsorption
- Humid chamber at 37 °C (*Protocol 2*)
- Filter paper wedges, length 2–4 cm

Method

A. *Pre-embedding labelling*

1. Perform the following steps instead of step 1 of the cell embedding protocol (*Protocol 3*).

2. Suspension cells: gently collect the cells by centrifugation (500 *g*, 10 min), discard the supernatant, resuspend the cells in a little PBG, and let them adsorb to the microplate for 30 min at room temperature.
3. Fix the immobilized cells (or monolayer cells grown in the wells) with fresh glutaraldehyde-formaldehyde mixture for 10 min.
4. For blocking, incubate with PBG-glycine or lysine for 5 min, and wash for 10 min with PBG; repeat this step.
5. Add the specific antibody to the wells and incubate for 1 h at 37 °C.
6. Remove the antibody, and wash the plate three times with PBG for 5 min each.
7. Incubate with the second labelled antibody for 1 h at 37 °C.
8. Remove the antibody, and wash twice with PBG for 5 min, and, finally, with buffer.
9. Post-fix the cells with 2.5% glutaraldehyde overnight, and continue with step 2 of *Protocol 3*.

B. *Post-embedding labelling of methacrylate-embedded specimens*

Perform on-section immunolabelling before the post-staining steps of *Protocol 3*.

1. Pipette the reagents onto a piece of Parafilm™ as described in *Protocol 2*.
2. For conditioning, place the grids with the section face down onto PBG-glycine or lysine for 10 min, and wash for 10 min with PBG.
3. Incubate the grid for 1h at 37 °C on a drop of the specific antibody.
4. Wash three times with PBG for 5 min each.
5. Float the grid for 1h onto a drop of the second antibody, and incubate at 37 °C.
6. Remove excess antibody using a filter paper wedge, and float the grid on two drops of PBG for 5 min each.
7. Wash twice with double-distilled water for 1 min each, and dry the grid using a filter paper wedge.

Note: If the surface of the ultrathin section is very smooth (e.g. embedded in Epon), not many determinants for immuno-labelling are exposed. To increase the available surface it may help to etch the section by treatment with 1% (w/v) sodium periodate in 0.5% (v/v) acetic acid for 20 min, followed by thorough washing with double-distilled water before immuno-labelling.

8. Limitations of TEM in cell culture

A severe limitation in the TEM detection of viruses and other agents is created by the need for particle concentrations higher than 10^6 physical particles per millilitre. The number of physical particles exceeds, however, the number of infectious particles by a factor of 10–10,000 fold, depending on the particular virus. This is partly due to the fact that, as judged from quantitative negative-staining experiments, the equivalent of only 3–5 μl can be adsorbed to the grid. The detection limit can be improved by two orders of magnification, using concentration by sedimentation and/or immuno-techniques. In view of this limitation, EM techniques have to be applied as one tool among others in a repertoire of further analytical techniques. In particular, when searching for contamination, one has to keep in mind that, due to the high detection limit, an absolute exclusion based on negative TEM observations is virtually impossible. To guarantee the quality and validity of the results obtained in EM, it is recommended to regularly assess the laboratory's performance, for example, by taking part in external quality assessment schemes and/or accrediting and certifying courses.

9. Safety precautions

Many chemicals used in electron microscopy are hazardous and should therefore be handled with appropriate care (*Table 6*). In addition, viruses, as well as the infected cell cultures, may cause safety problems, depending on the virus and its concentration (take note of specific information regarding the biohazard risk grouping). Also supposedly non-infected cell cultures should be handled with precautions. It is important to minimize the risk to employees, for example, by establishing 'clean areas' with access only for fixed virus specimens. Reasonable precautions involve storage, personal protection (e.g. laboratory coat, gloves, clean bench, vaccination), appropriate containment

Table 6. Hazardous chemicals in electron microscopy

Chemicals	Risk
Negative stains	Heavy metals: cumulative poisons; uranyl salts: radioactive in addition
Fixatives	Poisonous by ingestion, inhalation, or contact; osmium tetroxide is volatile
Liquid resins	May cause dermatological problems; some are carcinogenic, many flammable; adsorption into skin is increased when mixed with solvents; polymerized resin trimming dust: the small particles are thought to cause allergies

(e.g. of volatile chemicals), spillage protocols, and disposal procedures which accord with local safety regulations.

References

1. Murphy, F. A., Fauquet, C. M., Bishop, D. H. L., Ghabrial, S. A., Jarvis, A. W., Martelli, G. P., Mayo, M. A., and Summers, M. D. (1995). *Arch. Virol.* Suppl. **10.**
2. Aldrich, H. C., and Todd, W. J. (1986). *Ultrastructure techniques for microorganisms*. Plenum Press, New York.
3. Bozzola, J. J., and Russel, L. D. (1991). *Electron microscopy – principles and techniques for biologists*. Jones and Bartlett, Boston.
4. Robards, A. W., and Wilson, A. J. (1993). *Procedures in electron microscopy*. Wiley, New York.
5. Doane, F. W., and Anderson, N. (1987). *Electron microscopy in diagnostic virology: a practical guide and atlas*. Cambridge University Press, Cambridge.
6. Madeley, C. R., and Field, A. M. (1988). *Virus morphology* (2nd edn). Churchill Livingston, Edinburgh.
7. Nermut, M. V., and Stevens, A. C. (1989). *Animal virus structure*. Elsevier, Amsterdam.
8. Palmer, E. L., and Martin, M. L. (1988). *An atlas of mammalian viruses*. CRC Press, Boca Ratan, FL.
9. Mahy, B. W. J., and Kangro, H. O. (1996). *Virological methods manual*. Academic Press, London, San Diego, New York.
10. Harris, J. R. (1997). *Negative staining and cryoelectron microscopy*. Bios Scientific Publishers, Oxford.
11. Hayat, M. A., and Miller, S. E. (1990). *Negative staining*. McGraw-Hill, New York.
12. Hayat, M. A. (1989). *Principles and techniques of electron microscopy*. Macmillan Press, London.
13. Hyatt, A. D., and Eaton, B. T. (1993). *Immuno-gold electron microscopy in virus diagnosis and research*. CRC Press, Boca Ratan, FL.

6

Virus vaccines

R. JENNINGS and C. W. POTTER

1. Introduction

Any list of significant developments in the history of medicine would include the introduction of vaccines against infectious disease. History records epidemics and pandemics in earlier centuries which ravaged society and influenced or precipitated major events. The eighteenth century saw the dawning of the age of immunization, the nineteenth the wider introduction of vaccines against a variety of human and animal infections, through the work of many microbiologists. The present time has seen the eradication of smallpox, the predicted eradication of poliomyelitis and measles, and the control of many other infectious diseases. Vaccine development has been targeted to all virus infections which have a significant impact on human and animal health in terms of morbidity and mortality. The vaccine era is considered to have begun with the use of vaccinia virus to immunize against smallpox in the eighteenth century, and continued and expanded with the development of inactivated vaccines and live attenuated vaccines. Novel, modern vaccination strategies of the late-twentieth century are based on genetic engineering and the application of molecular biology, including the development of recombinant virus vaccines, replication-defective and DNA vaccines, together with the use of natural or fabricated accessory molecules and formulations for enhancing and directing the immune response. *Table 1* lists many of those virus vaccines presently commercially available for the immunization of man and animals, their form or type, and also indicates some virus vaccines currently under development and which may be generally available in the future.

Complementary to the wide and varied types and forms of virus vaccines has been the refining of safety and regulatory factors, which remain major concerns for all workers in the field, as novel technology and vaccine strategies come under consideration. Current concerns include the economic relevance of vaccine development; the regulatory requirements which govern vaccine production; standardization of vaccine constituents in terms of purity and reactive molecules; and the extensive clinical trials of vaccines for use in man, which must be carried out in order to justify their use, safety, and efficacy in target groups in different countries and in different age groups.

Table 1. Virus vaccines currently available or in clinical trials

Types of vaccine preparation	**Examples**
Inactivated whole virus	Hepatitis A; Poliomyelitis; Rabies; Japanese encephalitis; Tick-borne encephalitis; Influenza
Split virus	Influenza
Subunit/nucleic acid-free/± adjuvant	Influenza; Equine influenza; Epstein-Barr virus[a]; Respiratory syncytial virus[a]; Human immunodeficiency virus[a]; Parainfluenza[a]; Human cytomegalovirus[a]
Adjuvanted recombinant glycoprotein	Hepatitis B; Foot and mouth disease; Human immunodeficiency virus[a]; Herpes simplex virus[a]
Vectored virus antigen	Epstein-Barr virus[a]; Measles[a], Human papilloma virus[a]
DNA/plasmid	Human immunodeficiency virus[a]
Replication defective	Herpes simplex virus[a]
Live, attenuated reassortant	Rotavirus[a]; Parainfluenza[a]; Influenza[a]
Live attenuated	Poliomyelitis; Mumps; Measles; Varicella; Rubella; Yellow fever; Bovine respiratory syncytial virus[a]; Adenovirus; Argentine haemorrhagic fever; Louping ill; Influenza[a]; Human cytomegalovirus[a]; Dengue; Hepatitis A[a]

[a] In clinical trials.

The prelude to the development of a vaccine against a virus disease includes satisfactory answers to a number of important questions (*Protocol 1*). Firstly, epidemiological studies should establish that the infection warrants the development of a vaccine; the infection should normally be common, or severe and carry the threat of disability or death. This is important, since all vaccines carry the potential risk of side reactions, and these must be very significantly less in both frequency and severity than the disease itself. The agent of the infection must be known, and readily manipulated in the laboratory; methods of growth and assay should be available, or the agent should be accessible for manipulation using the techniques of molecular biology. The antigen(s) which stimulate the protective immune response should have been identified, and these antigen(s) should, ideally, be common to all serotypes of the particular virus, or at least show extensive cross-reactivity. A vaccine against the common cold is theoretically possible but with over a hundred non-crossreactive strains of rhinovirus, a vaccine is neither practical, nor does the relative mildness of the infection and the lack of adverse sequelae justify the development of such a vaccine. Studies of infection and immunity should have identified the nature of the immunity required to provide durable protection: immunity to some infections is based on the presence of circulating antibody, in other cases on cytotoxic T-cell responses, and in further instances both humoral and cell-mediated immune responses are required. Induction of T-cell memory is usually important. This information is crucial, since a vaccine must contain the appropriate antigen(s) and stimulate the correct immune response. The future may see the introduction of enhancement systems which can direct the immune response in an appropriate manner.

Protocol 1. Issues determining virus vaccine design, development, and use

1. Establish, through epidemiological studies, that the infection justifies development of a vaccine, and that the causative virus is sufficiently characterized.
2. Determine factors of viral pathogenicity, such as the existence of latent, persistent, or chronic infection, the presence of related viruses in the wild, viral pathogenic factors operative at cellular level, and transmission routes which may influence immunization strategy.
3. Host factors to be considered include severity and prevalence of the virus infection in immunocompetent and immunocompromised populations, the nature of the target population to be immunized, the existence of any pre-existing immunity, and the genetic constitution of the population.
4. Determine the nature and form of the virus antigen(s) for inclusion in the vaccine, and the type or form of vaccine to be used. Select those antigen(s) which induce protective and/or memory responses, are appropriately glycosylated, are stable, contain T-cell epitopes, cover antigenic variability, and exclude irrelevant or detrimental proteins or amino acid sequences.
5. Safety is of prime importance, and questions regarding the appropriateness of living or non-living preparations, the presence of virus DNA, host-derived proteins, residual live virus, or adventitious viruses in the preparation are addressed. The possibility of adverse reactions including oncogenicity, hypersensitivity, or other complications, appropriate attenuation, reversion of virus to virulence or virus recombination, and the suitability of the substrate used to raise the vaccine are investigated.
6. Generation of the most relevant and appropriate immune response(s) will depend on vaccine antigen presentation and processing, the adjuvants, vectors, or delivery systems used, the route of vaccine administration, and the immunizing dose and regime employed.
7. Ensure that production methods meet the required safety standards, and that vaccine batches are carefully and fully characterized.

Protocol 1. *Continued*

8. Phase I and phase II clinical trials should precede phase III clinical trials. These determine vaccine safety and efficacy, measured using the most appropriate assessment systems and conditions.
9. Other issues requiring deliberation include overall immunization policy, particularly issues of disease eradication, combination vaccine strategies, vaccine distribution, and cost.

Once the above requirements have been met, and specific virus or host factors that may impinge on vaccine development and usage, such as possible virus integration into host cell chromosomes, virus latency, host autoimmune responses, and host genetic make-up are taken into account, further development, clinical studies and evaluation of the final vaccine formulation can be undertaken prior to commercial release. Given the above, vaccine manufacturers and developers have prepared, or have under study, a wide range of vaccines against virus diseases of man and animals. For extensive detail on individual virus vaccines, the reader is referred to the comprehensive volume edited by Plotkin and Mortimer (1).

2. Non-living virus vaccines

Despite the advances in the new technologies of molecular biology and genetic engineering, and the focusing of these technologies on the development of novel strategies for vaccines, a number of virus vaccines developed and evaluated in the past remain commercially available and in use at the present time. Vaccines consisting of whole inactivated virus are available for use against infections such as poliomyelitis, tick-borne encephalitis, Japanese encephalitis, rabies, influenza, and hepatitis A, while split and subunit virus vaccines are in widespread use against influenza (1). These conventional strategies for virus vaccines remain relevant even as researchers explore novel avenues for control of other virus diseases. Whole, inactivated, or subunit antigen vaccines have been considered, and indeed have reached clinical trial stage, for viruses such as the type 1 human immunodeficiency virus (HIV-1), the human cytomegalovirus (CMV), human parainfluenza virus, Epstein-Barr virus (EBV), respiratory syncytial virus (RSV), and others (*Table 1*). Nevertheless, modern technology has made an important impact on the design of non-living virus vaccines, and many virus proteins, glycoproteins, or specified regions or epitopes of these antigens, can be prepared to high concentration and purity by recombinant DNA technology. To date this has resulted in licensing of a hepatitis B surface antigen (HBsAg) vaccine for use in man, and a foot-and-mouth disease (FMD) vaccine for use in animals.

2.1 Non-living vaccines produced by conventional methods

Once the virus strains to be used have been carefully characterized with regard to their passage history and properties, the preparation of whole or split subunit virus vaccines involves propagation of the selected virus strains to high titre on a well-defined, acceptable, and licensed substrate, usually some form of cell culture such as Vero cells or human diploid fibroblast cells, or for influenza virus vaccines, pathogen-free embryonated eggs (*Protocol 2*). Japanese encephalitis virus vaccine, however, is produced from infected mouse brains. Virus, usually obtained in high titre as cell-free fluid, is subjected to a number of purification and concentration steps, which may include filtration, ultracentrifugation, chromatography, and precipitation. These processes remove cellular proteins, cellular nucleic acids, and other undesirable, non-virus materials such as residual antibiotics from the cell-culture medium. Inactivation of the purified pools of virus is usually by treatment with formaldehyde or β-propiolactone under carefully defined conditions.

Protocol 2. Production of non-living virus vaccines by conventional procedures

1. Grow the strain of virus known to carry the required antigens, rigorously characterized and selected by plaque purification procedures, in bulk to high titre using a defined substrate under defined conditions. A fully characterized cell line or other acceptable substrate authorized for propagation of viruses for use in vaccines is employed.
2. Harvest the virus at its replication peak. Large volume pools are prepared and subjected to a purification process, usually including differential centrifugation and filtration, to remove unwanted, non-virus material.
3. Inactivate the virus pool using β-propiolactone or formaldehyde under the appropriate, defined conditions.
4. Standardize and screen the inactivated, whole virus vaccine with respect to total protein content, antigen content and specificity, absence of living virus, lack of extraneous substances, sterility, and endotoxin content.
5. A split or subunit vaccine can be prepared following release of surface glycoprotein antigens (usually considered relevant to protection) from inactivated whole virus by treatment with detergents, such as Triton or Empigen, or with ether-Tween. Alternatively, the relevant surface antigens may be purified from split virus preparations to form a subunit vaccine.

Protocol 2. *Continued*

6. Separation of the required virus antigens, usually glycoproteins, from unwanted virus or cellular proteins, and unwanted virus or cellular nucleic acid, is achieved by procedures which include dialysis, filtration, and centrifugation.
7. Subunit vaccine preparations are standardized and screened for antigen content and specificity, purity, total protein content, lack of living virus, sterility, and endotoxin content.
8. Test both non-living whole or subunit virus vaccines for toxicity, pyrogenicity, potency, antigenicity, and immunogenicity. An adjuvant, which for vaccines for human use can only be aluminium salts, may be incorporated.
9. Add preservative, dispense the vaccine as required into suitable containers, which may be syringes ready for injection.
10. Standardize and screen samples from each aliquoted batch as described above.

At this stage, the inactivated whole virus preparation is standardized with respect to protein, virus antigen and endotoxin content, specificity, and potency, and checked for sterility, absence of living virus, pyrogenicity, antigenicity, and immunogenicity. These standardization procedures may include both *in vitro* and *in vivo* tests. For example, inactivated whole poliovirus vaccine is standardized *in vitro* for D antigen content, and *in vivo* for its ability to produce antibody responses in rodents, in comparison to a World Health Organization reference standard of known immunogenicity in infants. The specific virus antigen content of hepatitis A vaccine is determined by a standardized enzyme-linked immunosorbent assay (ELISA).

Split or subunit virus vaccines, such as some influenza virus vaccines, can be derived from whole virus preparations obtained and purified as described above, and subsequently treated with detergents or organic solvents to solubilize the surface antigens, often glycoproteins, regarded as important in inducing protection. These virus antigens are separated from irrelevant or unwanted virus proteins, virus nucleic acid, and envelope lipids, by procedures which may include dialysis, ultracentrifugation, and filtration. The resulting split virus or subunit preparation is standardized and tested by both *in vitro* and *in vivo* tests as described above. The final vaccine product will contain the required virus antigens, which in the case of the influenza virus vaccine consists of aggregated monomers of the haemagglutinin (HA) and neuraminidase (NA) glycoproteins. Whole virus or subunit non-living vaccines may be absorbed onto aluminium hydroxide gel acting as an adjuvant to enhance immune responses, and also to reduce toxicity. Preservative, usually thiomersal, is added, and the vaccine dispensed into suitable containers for storage, transportation, and use as required.

2.2 Non-living vaccines produced by recombinant DNA technology

Although it is a subunit vaccine preparation consisting solely of the aggregated virus surface protein (HBsAg), the hepatitis B vaccine is prepared by modern recombinant DNA technology (2), rather than by the conventional procedures described above for other non-living vaccines, and represents the first virus vaccine derived using such technology to become commercially available for use in humans. In brief, the gene coding for the HBsAg is incorporated into a plasmid, together with other required sequences permitting functional expression of the antigen, and transfected into either yeast (*Saccharomyces cerevisiae*) or mammalian (Chinese hamster ovary) host cells. Faithful translation of the HBsAg message, and expression of this protein to high level, occurs in these cells, and in addition the protein adopts the correct secondary and tertiary structures. The antigen is then extracted from disrupted yeast cells and, using a series of chromatographic separation steps, obtained as a highly purified product in high concentration. Vaccine batch characterization for impurities, protein content, antigen specificity, potency, and antigenicity are all carried out by standard procedures.

Recombinant DNA technology is currently being intensively applied to the development of novel vaccines against a number of human and animal virus diseases, including those caused by the HIV-1, HSV, influenza, and foot-and-mouth disease viruses (3). This technology allows production not just of highly purified virus proteins or glycoproteins of importance in protection, but also of epitopes or short peptides, such as the V3 loop of the HIV gp120 protein, and short peptide sequences of influenza or HSV glycoproteins believed to be important in inducing protection. Modern technology can also allow the synthesis of such peptides as the V3 loop region of HIV gp120. Expression systems for recombinant virus proteins and glycoproteins include yeasts, mammalian cells, and insect cells. Such proteins and peptides can be fused to, or genetically engineered to be linked with, other proteins such as cytokines, or conjugated with or incorporated into various adjuvants, carriers, or delivery systems to enhance their immunogenic potential or to target them to particular cells or tissues.

Although non-living, conventional whole, split, or subunit virus vaccines are still extensively used as licensed, commercially available products, there are deficiencies associated with some of these preparations. There is therefore, considerable research into developing improved, replacement vaccines, or modifying existing vaccine preparations to enhance their immunogenicity or reduce their reactogenicity, using the new technologies, novel adjuvants, and immunomodulators.

3. Live, attenuated virus vaccines

3.1 Existing live attenuated vaccines

Although inactivated virus vaccines, either whole or split/sub-unit, may be perceived to be the easiest, cheapest, and safest type of vaccine for immunization against virus diseases, these are not always either satisfactory or the best immunogens. Thus, inactivated influenza vaccines have been used for many years, although much evidence indicates that live vaccines induce a more solid immunity, and a degree of protection against heterologous strains of virus. Inactivated measles vaccines induce only short term immunity, which may be followed by atypical measles infection when immunity has waned. Inactivated poliomyelitis vaccine protects against clinical disease, but does not protect against carriage of the virus, and could never satisfy the ambition for eradication of this disease. These few examples of the limitations of inactivated vaccines, coupled with the accepted belief that live infection provides a more solid immunity than that induced by inactivated vaccines, has been the impetus for the development of live attenuated virus vaccines against many virus infections (1).

The disease against which a live attenuated vaccine is to be produced should ideally be caused by a virus which can be easily grown, manipulated, and purified in the laboratory. The selection of the virus strain as the starting point for the development of attenuated virus vaccines has been arbitrary: researchers have isolated virus in the laboratory from a clinical case of the disease, and attempts to attenuate the strain have proceeded empirically, if this was indeed the original intention of the scientists involved. A selected strain does not usually possess any property marking it as appropriate for vaccine development. In most cases, no attempt was made to plaque-purify the virus prior to the attenuation process.

The methods used to attenuate virus for vaccine use (*Protocol 3*) are as varied as the number of virus vaccines available, and no method has been the same for any two viruses. However, two basic principles run through the process. Firstly, attenuation of virus has been by serial passage in clinically inappropriate cells. Secondly, passage has commonly been carried out at low temperatures. Thus, attenuation of the bovine respiratory syncytial virus, human herpes simplex virus type 1, rubella, and measles viruses, amongst others, has been achieved by passage at low temperatures of 28–35 °C. A variety of cell types, in addition to intact animals, have been used for serial passage to produce attenuated strains, including kidney, lung, and fibroblast cells from chicks, mice, hamsters, guinea pigs, and man, and in many cases passage has been through a number of different cell types:

- polioviruses were attenuated by rapid passage of high titre virus in monkey kidney cells
- a commonly used measles vaccine was attenuated by passage 24 times in

human kidney cells, followed by 28 passages in human amnion cells, and 6 in chick embryo cells

- the attenuated Jeryl Lynn strain of mumps virus was derived by passage in embryonated eggs and chick embryo fibroblasts
- rubella vaccines have been made by attenuation in duck embryo cells, and in dog and rabbit kidney cells, while the commonly used rubella virus vaccine strain, RA 27/3, was attenuated by eight passages in human diploid fibroblasts at 37 °C, followed by seven passages at 30 °C.

The attenuation process for other viruses is as variable and as singular. Finally, the number of passages needed for satisfactory attenuation has varied. For RA 27/3 rubella vaccine, 15 laboratory passages in tissue culture sufficed; for one measles vaccine, 58 passages in three different cell lines was used; and for the 17D strain of yellow fever virus vaccine, over 200 passages were made in mouse brain, rhesus monkey and mouse cells, and finally chick embryonic cells, to produce the current attenuated vaccine.

Protocol 3. Live, attenuated virus vaccines

1. Select a virus isolate from a human case, and passage at low temperature in a tissue culture system distinct from the cells which are the target cells of natural virus infection.
2. Serially passage the virus numerous times in one or more cell systems until attenuation has been achieved, as shown empirically. This is established by inoculating appropriate cells, and demonstrating a loss or reduction in virus infectivity, or by inoculation into a susceptible animal species to demonstrate loss of virulence in comparison to the wild-type virus.
3. Plaque purify the virus in an approved cell substrate.
4. Amplify the attenuated virus strain by growth in an approved cell substrate. These are usually established and well characterized normal cells, free of adventitious agents, and having a stable and normal chromosome complement. This process takes place in approved and licensed premises, where all stages of production are rigorously controlled.
5. Inoculate the attenuated virus vaccine strain into a small group of individuals who are monitored for reactions. Only if the virus regularly and consistently produces little or no clinical reaction, and certainly significantly less reactions than those produced by the natural disease, can the programme go forward.

The problem of antigenic variability is seen for many viruses, such as influenza and HIV-1. Whilst many viruses are stable and a single vaccine immunizes against the disease, rapidly mutating RNA viruses often show variability in their major surface glycoproteins. In the case of influenza, new vaccines containing the relevant glycoproteins are required each year, and the process of attenuation by passage is too long and unpredictable to be of any value. For HIV-1, the generation of many antigenic variants during replication in humans, due to mutations in regions of the nucleic acid coding for the gp160 surface antigen of the virus, has posed problems for the preparation of all types of vaccines against this virus.

Contamination of live, attenuated vaccines by adventitious agents has been reported, and this remains individual to the vaccine. Thus, the 17D strain of yellow fever vaccine was found to be contaminated with avian leukosis virus, and was freed of this agent to produce the present vaccine; adenovirus vaccines were contaminated by SV40 derived from the monkey kidney cells originally used - indeed, SV40 is essential for replication of adenoviruses in these cells - and are now prepared in human embryo kidney cells.

A putative attenuated virus, which may show diminished replication *in vitro* or low-temperature dependence, is then often tested in an animal model. For many infections, animal models are available:

- rhesus monkeys are very susceptible to poliovirus, indeed, more susceptible than man, and attenuated polioviruses have reduced neuropathology in this model
- live influenza virus vaccines have been extensively tested in ferrets and can be shown to be attenuated
- HSV vaccines can be tested in mice and guinea pigs

Successful vaccines need to be demonstrably attenuated compared to wild-type virus, and all vaccines currently available that have been tested in animals have this property. However, a suitable animal model is not available for all viruses. Thus, there is no animal model for adenovirus infection, and live adenovirus vaccine types 3, 7, and 21 were given in phase I studies to volunteers in an encapsulated form where they cause intestinal infection followed by immunity, but do not cause respiratory disease. Rubella vaccines have all been tested in a limited number of human volunteers prior to wider study; while Argentine haemorrhagic fever vaccine was also tested in small numbers of volunteers prior to large-scale clinical trials.

Prior to studies in volunteers, a candidate live, attenuated vaccine virus must be cloned by plaque purification or limiting dilution techniques one or more times, cultured in a licensed cell substrate, and shown to be free of contaminating adventitious agents (*Protocol 3*). This purification is important, as mentioned earlier for inactivated vaccines, since mistakes have been made in the past, and recognized in hindsight: early poliovirus vaccine preparations were contaminated with SV40 virus present in monkey kidney cells, as were the adenovirus vaccines, while the louping ill vaccine used for sheep was contaminated with the scrapie agent, and this later caused disease in a large number of vaccinated animals. A number of cell lines have been licensed as substrates for the production of vaccines, including human embryonic lung fibroblasts, human embryonic skin fibroblasts, chick kidney cells acquired from leukosis-free eggs, and certain monkey kidney cell lines: continuous and tumour cell lines are not permitted. The animals from which these cell substrates are derived must be shown to be pathogen-free, and the cells must be shown to be genetically stable. Cloned attenuated virus is amplified in an appropriate and approved cell substrate to give vaccine lots for human investigation. This relatively simple statement embodies a wealth of legislation to ensure purity and safety at every step. The process must be carried out on licensed premises using cloned attenuated virus, and on permitted cells; samples are taken at every stage in the operation for quality control; and the rules that govern the whole procedure and the eventual handling of the material is governed by rigorous legislation.

Volunteer studies with putative attenuated virus vaccines are usually carried out when all the above criteria have been fulfilled, including attenuation of the virus, demonstration of attenuation in an animal model where possible, cloning of the virus, and the production of vaccine virus lots carried out under strict regulatory control (*Protocol 3*). At this point the virus may be given to a small group of volunteers. This is done to prove safety and to demonstrate attenuation for humans, or to ensure that if some reactions do occur they are permissible and significantly less than those induced by the natural disease. This procedure has been carried out with virtually all live virus vaccines so far developed, including poliomyelitis, measles, mumps, rubella, flaviviruses, arenaviruses, and others. It is at this point that problems associated with the use of live vaccines are often realized:

- some of the earlier rubella vaccines, although attenuated, were found to induce an unsatisfactory level of immunity and were discontinued
- one of the mumps vaccines originally introduced was withdrawn, since it caused meningitis in a small number of volunteers
- a measles virus vaccine caused reactions which could be controlled by addition of IgG antibody
- although a dengue 1 attenuated vaccine, now under development, appears to be a safe vaccine, a dengue 2 live, attenuated vaccine has produced some adverse reactions and may require further attenuation
- many of the putative attenuated influenza vaccines were shown to be genetically unstable, often reverting to virulence, and were considered unsuitable

Should a live, attenuated vaccine virus satisfy safety requirements in small groups, larger groups may be immunized in phase II clinical trials to test safety further, and to demonstrate adequate and durable immunity and protection against challenge infection; such studies have been carried out with all existing licensed live attenuated vaccines, and are intended for other such vaccines in preparation.

3.2 Live, attenuated vaccines based on reassortment

One specific approach to obtain attenuated virus vaccines for human use is through the production of reassortant viruses, and this strategy has been employed especially for the development of live, attenuated influenza virus and rotavirus vaccines. For influenza viruses with intrinsic and continuous antigenic variability, the preparation of a reassortant virus for use in vaccine production bearing the surface glycoproteins, derived from current, epidemiologically relevant strains and a master strain supplying the characteristics of high yield and rapid growth, as well as the epidemiologically less important internal proteins, is the only available strategy for the realistic development of a live, attenuated vaccine against influenza within the relatively short time available. Cold-adapted A/Ann Arbor/6/60 virus, when reassorted with wild-type virus, has produced reliably attenuated strains each year for the past thirty years: reassortant strains based on cold-adapted B/Ann Arbor/1/66 have been equally successful. Both the influenza A and B reassortants contain large numbers of mutations in various genes; have reduced pathogenicity for mice and ferrets; are stable when inoculated into volunteers; induce local and systemic immune responses, and provide immunity against challenge virus infection (4). Reassortants derived from both the influenza A and B cold-adapted strains have already been administered to large numbers of individuals in high-risk groups, and live attenuated influenza virus vaccines based on this strategy are being developed by Avion (Mountain View, California), and are currently in phase III clinical studies in healthy adults and children, with the expectation that they will receive general licence when these studies are completed in 1999. A further strategy currently being applied for the design of live, attenuated influenza virus vaccines is the introduction of cDNA-derived RNA into influenza viruses by transfection of ribonucleoprotein complexes into virus-infected cells. This use of reverse genetics has allowed the exchange of six genes between different influenza virus strains, and may permit the development of deletion mutants which cannot revert to virulence, representing a novel means for obtaining strains of these viruses suitable for live, attenuated vaccines (5).

Influenza is not the only virus disease for which vaccines can be developed by a reassortment strategy; the principle has also been applied to the development of a tetravalent human rotavirus vaccine. Rotaviruses are responsible for up to three million deaths annually, and a vaccine against this disease has been sought for many years. Human rotaviruses are pathogenic, but the rhesus monkey rotaviruses are naturally attenuated and immunogenic for humans.

Reassortant viruses from rhesus and human rotaviruses, containing the human rotavirus genes that specify the virus surface antigens responsible for stimulating the protective antibody response, have been developed; this reassortant virus vaccine given orally to infants in three doses at age two, three, and five months has been shown in recent trials to be safe, and to protect against natural disease (6). It is anticipated that this vaccine will be licensed and available in the near future.

4. Delivery of vaccine antigens through vector technology and live, recombinant micro-organisms.

A number of micro-organisms can be considered as vectors for delivery of the host of virus antigens relevant for protection (7). These micro-organisms, both bacteria and viruses, can act as vectors by virtue of characteristics such as their known avirulence or attenuated virulence, safety and stability in the host, persistence in the host, ease of administration, immunogenic potential, ease of generating recombinants, and possession of a large genome. Certain viruses with known safety records in man, with fully characterized, large genomes, and well-marked attenuation characteristics, such as vaccinia virus (8), and recently the Oka strain of varicella zoster virus (9), may be appropriate for this purpose. Genes from a large number of viruses have already been genetically engineered by recombinant DNA technology into vaccinia virus. Advantages of this technology are induction of a broad immune response to the expressed foreign protein through the live nature of the vector virus; amplification of the immune response in the host through replication of the vector virus, and hence the expressed foreign protein; and delivery of the foreign protein into the cell cytosol, allowing antigen processing to elicit class I MHC-restricted immune responses and cytotoxic T lymphocyte (CTL) activity.

If the vector virus genome is large enough, it may be possible to incorporate genes from several viruses, enabling vaccination against several diseases to take place using a single vector. Use of a micro-organism as a vector may, however, entail genetic modification of the micro-organism to eliminate or nullify genes associated with unwanted pathogenicity factors or restriction of immune responses. One recent advance has been the development of a highly attenuated, replication-deficient, modified vaccinia virus, Ankara (MVA), which may overcome some of the safety problems associated with the use of vaccinia virus as a vector (10).

The detailed technology for construction of recombinant viruses and the expression of foreign genes by these micro-organisms is described in detail in many reviews, but involves identification in the vector genome of regions which can either be deleted or interrupted by insertion of the foreign gene(s) (*Protocol 4*). These are normally non-essential regions for vector replication.

The foreign gene of interest, which may have to be modified by site-directed mutagenesis to eliminate unwanted termination or other signals, is cloned into a plasmid along with flanking promoter and stop codon sequences to direct expression, and vector sequences that will permit insertion of the foreign gene by homologous recombination into the vector (together with its expression sequences). The plasmids, after validation of their construction by DNA sequencing, and the virus vector are then introduced, the former by transfection, into an appropriate cell culture system to allow recombination to occur. Recombinant progeny viruses are separated and identified by a distinguishing signal, often included in the plasmid construct (e.g. β-galactosidase), by the polymerase chain reaction (PCR), or by limiting dilution cloning.

Protocol 4. Vector technology and the development of live, recombinant virus vaccines

1. Select vector virus strain with known safety and immunogenicity record in the host, for example certain vaccinia virus strains, herpesviruses, or adenoviruses.
2. Ideally, no evidence of oncogenicity or teratogenicity is associated with the vector virus genome, although, if present, the genes responsible can be deleted.
3. The vector will possess a relatively large, fully mapped genome, with readily accessible insertion sites and/or deletable, non-essential genes.
4. The DNA chosen for recombination with the vector genome will code for a virus protein(s) known to be protective, immunogenic, or to be an important virulence factor.
5. Coding sequences for the chosen gene(s), together with required promoters, are generated by PCR amplification and incorporated into plasmids using restriction endonucleases and ligation techniques.
6. Plasmids can be constructed and transfected into appropriate vector-infected cell lines to allow recombination to occur.
7. Growth of progeny virus is detected by plaque formation, often through a pre-determined distinctive signal, and the desired progeny recombinant virus selected, uncontaminated by wild-type vector virus, by limiting dilution cloning.
8. The presence or expression of the inserted gene sequence(s) should not modify properties of the vector virus such as growth rate, interaction with the host, attenuation characteristics, or immunogenicity.
9. The introduced gene(s) will be expressed to high level, be genetically stable, and not be integrated into the host cell genome.

10. Presence of the inserted gene(s) in the vector is determined by Southern blot analysis, and gene expression, usually at the surface of cells infected by the recombinant virus, assessed by immunoblot analysis.

Isolation of the recombinant virus, and verification of the presence of the foreign gene sequences, using a technique such as Southern blot analysis, are followed by further characterization of the recombinant vector virus to demonstrate that the foreign gene is satisfactorily expressed, usually at the surface of cells infected by the recombinant, and also as part of the virion itself. In addition, tests are carried out to ensure that the attenuation properties, growth capability, immunogenicity, and cellular interactions of the recombinant have not been seriously perturbed by the genome changes.

A variety of both DNA and RNA viruses have been used for the construction of recombinants expressing foreign virus antigens, including vaccinia virus, the prototype vector virus, adenoviruses, adeno-associated viruses, herpesviruses, and picornaviruses. Apart from the use of vaccinia virus as a live, attenuated, replicating virus vector, most attention has been focused on the adenoviruses, a number of which have a long history of use as live attenuated vaccines, delivered by the oral route for the prevention of respiratory infections in certain target groups. The capability of adenoviruses to act as vectors delivering expressed foreign genes to target mucosal surfaces in the respiratory tract or the gut, thereby stimulating mucosal immune responses, represents a particularly useful property of this group of viruses. Adenoviruses grow well in cell culture, while genes responsible for cell transformation by these agents in animal model systems can be deleted. The ability of some adenoviruses to persist in cell culture systems, and probably in humans, represents a two-edged sword; prolonged antigenic stimulation of a foreign protein expressed by an adenovirus vector could be immunologically beneficial, but any long-term effects of persistence on factors such as stability of the vector have not been demonstrated.

Other strategies under intensive study for virus vaccine antigen delivery through vector technology include the use of live, non-replicating vector viruses, such as the avipoxviruses, in particular canarypoxvirus, which due to host restriction is incapable of replication in human cells, and the insertion of virus genes into live attenuated bacteria such as *Mycobacterium bovis* (BCG) and genetically engineered auxotrophic mutant strains of *Salmonella typhi* and *Salmonella typhimurium*. These bacteria, with low or considerably reduced virulence, have been constructed to carry and express such foreign virus genes as hepatitis B pre-S or core antigens, influenza virus haemagglutinin or nucleoprotein antigens, herpes simplex virus glycoprotein D, and the rotavirus VP7 antigen.

As with most other new technologies under development in the virus vaccine field, delivery of virus antigens relevant to protection and control of virus disease, through live, recombinant vector micro-organisms has progressed to the stage of clinical trials. However, with a number of problem issues pertaining to the use of these live vectors, their acceptance as a viable strategy for delivery of virus vaccine antigens remains unresolved.

5. DNA vaccines

Reports that inoculation of animals with plasmid DNA containing virus genes resulted in protein expression and immune responses alerted researchers to the potential of such plasmids as vehicles for immunization. The literature now records that plasmids incorporating DNA sequences for proteins or glycoproteins of influenza virus, lymphochoriomeningitis virus, rotavirus, rabies virus, retroviruses, papillomaviruses, bovine herpesvirus type 1, human and simian immunodeficiency viruses, hepatitis B and C viruses and herpes simplex virus type 1 have been inoculated into one or more animal species including mice, monkeys, ferrets, cows, guinea pigs, and chickens to induce a range of immune responses. In many cases, these responses have been shown to protect against challenge virus infection. Indeed, the list of potential DNA vaccines extend beyond the limits of virology to bacterial and protozoan diseases, and to allergy and malignant diseases. However, many studies have been carried out in mice using plasmids that contain DNA sequences for influenza virus proteins (11). Several reviewers have listed the advantages of DNA vaccines, and contend advantages over conventional vaccines:

- DNA vaccines are reported to induce an immunity which is at least as effective as live virus infection, and more effective than inactivated vaccines
- by incorporating into plasmids only those virus DNA sequences which relate to the immune response, the vaccine can be made specific
- DNA vaccines can induce serum antibody, CTL responses, and local IgA antibody, and these responses are long-lived
- encoded virus sequences can be selected to induce B-cell and/or T-cell responses
- no reversion to pathogenicity can occur
- immune responses can be induced in the presence of maternal antibody and in the very young, where the response to conventional vaccines is limited

DNA vaccines are easy to prepare and purify, are stable and relatively inexpensive, and only small quantities of DNA are required. In addition, incorporated DNA could include sequences for a variety of viruses, to form a polyvalent vaccine against a spectrum of infectious diseases. However, much remains to be understood about DNA vaccines. Safety factors have not been fully investigated, and the underlying mechanism(s) that elicit the immune responses observed have not been fully determined: cautious optimism for DNA vaccines is the current sentiment (12). A strategy for the preparation of a virus DNA vaccine is described in *Protocol 5*.

Protocol 5. Strategy for the preparation of a virus DNA vaccine

1. Identify and isolate the gene sequence encoding the protein(s) which induces the immune response required. For RNA viruses, convert to DNA using standard molecular biology techniques (reverse transcription).
2. Insert the gene, together with any required introns, downstream of an appropriate eukaryotic promoter. Include a reporter gene such as *lac Z*.
3. Verify the fidelity of the inserted DNA by sequencing.
4. Transfect the construct into suitable mammalian cell lines, and verify protein expression to confirm the ability of the plasmid to direct protein synthesis correctly.
5. Select a suitable animal model, and inoculate the plasmid into the appropriate tissue (skin, muscle, etc.). Verify translation of protein.
6. Monitor antibody and/or T cell responses against virus antigen.
7. Test biopsy material to ensure no integration of DNA sequence into host cell chromosomes. Test blood samples from animals for anti-DNA autoantibodies.

There is evidence that immune responses to virus sequences in DNA plasmid vaccines are related to features of the plasmid construct, and some authors have suggested that the constructs may need to vary according to the nature of the virus insert(s), the species or tissue which is to receive the vaccine, and the immune response required. However, a DNA vaccine must contain an origin of replication to promote a high yield of plasmid DNA in *E. coli*; antibiotic selection markers to select plasmid-containing bacteria; a strong virus promoter, for example, the human CMV immediate early promoter, or the less strong promoter sequences from SV40 or Rous sarcoma virus; and a transcription termination sequence of SV40 or human or bovine growth hormone. Constructs may include a signal sequence, to facilitate secretion of translated protein, and an intron, since translation in mammalian cells may depend on this feature. The absence of an origin of replication which is functional in eukaryotic cells ensures that plasmids do not replicate, and are unlikely to integrate into host cell chromosomes. The virus DNA sequence included can be selected to induce humoral or cellular aspects of the immune response, and can be amplified by inclusion of sequences coding for appropriate cytokines. The inclusion of sequences coding for CD80 or CD86 can result in expression of these proteins at the surface of infected cells to promote T-cell receptor-MHC-peptide sequence interactions. Variations in DNA constructs abound, and much work remains to optimize these constructs, particularly as there is some evidence that elements in the plasmid backbone may be involved in the immunogenicity of these preparations.

The strength of the immune response to DNA vaccines is directly related to the extent of protein translation, and is influenced by plasmid design and route of plasmid inoculation. DNA vaccines can be given in saline, complexed with lipid, or encapsulated in particulate carriers. They can be inoculated intramuscularly or subcutaneously, or given intradermally by syringe or using a gene gun, either as aerosols or dried on to gold beads. This last method is favoured by many researchers. Both intramuscular inoculation and gene gun administration induce long term gene expression and immune response(s) comparable to that following virus infection. The uptake of DNA is optimal in muscle cells, and this uptake can be improved by pre-treating the tissues with hypertonic sucrose solution, snake venom, or the anaesthetic bupivicaine.

Translation of protein from DNA plasmids that contain sequences for a wide range of virus proteins is followed by intracellular translocation and conformational events similar to those following natural infection, with subsequent immune responses. It is presumed that expressed proteins are processed through the cell by pathways similar to those utilized by peptides derived as a consequence of natural virus infection, and responses can also be seen following oral or intranasal administration of DNA vaccines, when IgA responses have also been reported. All arms of the immune response can be stimulated, and this has been shown for many DNA vaccines; however, much of the mechanism is assumed, and much remains to be demonstrated (11). Thus it is difficult to rationalize the efficacy of intramuscular immunization of DNA vaccines with the paucity of MHC Class II molecules on muscle cells and the scarcity of antigen-presenting cells (APC) in this tissue. The antibody response to DNA vaccines is slower than that following infection or immunization with protein vaccines, and immune responses can be induced in newborn animals that are not seen following other types of immunization. Clearly, the immune response following inoculation of DNA vaccine has some singular features. However, immune responses to DNA vaccination in numerous systems have been shown to equate with immunity to challenge virus infection. It remains to be seen if translation of these observations to man can be made, whilst further progress in the development of DNA vaccines may depend on a better understanding of the immune mechanisms involved.

At present, immunization with DNA vaccines has both practical limitations and theoretical problems (13). Firstly, no DNA vaccine has yet been shown to be superior to the best existing vaccines, and the future of plasmid vaccines may be limited to infections where no conventional vaccines are available. Secondly, intramuscular inoculation of DNA vaccines can induce a mild inflammatory response in unprimed animals, and a distinct myositis in primed animals; this response is clearly antigen-specific, as inoculation with control plasmid has no effect. Thus, as with all injected vaccines, some reaction at the site of inoculation is seen. Thirdly, investigators have been concerned with the strong and sustained immune responses that follow plasmid DNA immunization; this, in conjunction with antigen persistence, could lead to immunocomplex disease, autoimmune reactions, tolerance, or anaphylaxis. No thorough investigation has been published to confirm or refute the above possibilities, and no evidence of autoimmune disease or anti-DNA antibodies has been found in the few studies which have addressed this question. Further work is needed to satisfy safety requirements on this count. Finally, concern has been expressed that inoculation of DNA vaccines may lead to the incorporation of DNA sequences into host chromosomes, which could cause insertional mutagenesis, leading to cellular transformation and malignant changes. Although the structure of the plasmids used theoretically precludes integration, and no evidence of integration has been found in studies where the question has been addressed, more research is required to clarify this point.

6. Replication-deficient viruses as vaccines

The use of replication-deficient viruses represents a novel strategy for virus vaccines, which can be perceived as an approach allowing some of the advantages of live, attenuated vaccines to be achieved whilst avoiding drawbacks inherently associated with such vaccines. A genetically engineered, live, replication-deficient virus designed for use as a vaccine has been generated from a type 1 herpes simplex virus (HSV-1) mutant lacking the glycoprotein H (gH) gene, essential for the replication of this virus (14). The gH gene was engineered into F6 Vero cells, allowing these to act as a complementing cell line such that after transfection of these cells with the gH^- HSV-1 mutant, progeny virus bearing the full complement of proteins but lacking the gH gene could be obtained. This virus, and also a more recently developed gH^- replication-deficient HSV-2 virus (15), are highly immunogenic in mice and guinea-pigs, and, following extensive safety and protection studies in these animals, the latter virus is currently undergoing phase I clinical trials. Other replication-deficient HSV-1 mutants have been described as vaccine candidates, including virions lacking the gene sequences for the virus host shut-off protein and the ICP8 protein, the major DNA binding protein of HSV.

Development of these replication-deficient viruses commences with selection of an appropriate, well characterized parental virus strain, preferably having low virulence properties, and as for other virus vaccines, selection of a cell substrate approved for the manufacture of biologicals for use in humans. After construction of a plasmid expression vector containing the virus gene to be deleted, this is transfected and established in the appropriate cell line (*Protocol 6*). A promoter is required to drive expression of the gene in the cell, and the plasmid should be constructed to allow no overlap of homologous sequences between the virus and the cell line, so that wild-type recombinants cannot arise. Cells carrying the virus gene are selected by use of a marker gene, usually for resistance to the action of an antibiotic, and limiting dilution cloning.

Protocol 6. A strategy for preparation of live, replication-deficient DNA viruses for use as vaccines

1. Select virus strain with no or low neurovirulence characteristics, and an appropriate cell substrate approved for manufacture of biologicals for human use.
2. Using appropriate recombinant DNA techniques, construct a plasmid expression vector containing an intact copy of the virus gene to be deleted, for transfection and establishment in the complementing cell line.
3. After transfection of the plasmid expression vector (which will also carry a selectable marker gene, usually for antibiotic resistance), isolate a cell line carrying the virus gene by limiting dilution cloning.
4. Construct a plasmid with DNA sequences flanking the virus gene sequence(s) to be deleted, together with a selectable marker sequence, usually *lac Z*, and an appropriate promoter, usually the CMV immediate early promoter.
5. Transfect this plasmid and purified virus DNA into an appropriate cell line.
6. Recombinant virus carrying the *lac Z* gene in place of the deleted virus gene is recognized by colour-based plaque assay, and purified by limiting dilution techniques.
7. The *lac Z* gene can subsequently be removed from the virus deletion mutant by a further recombination step, using a second plasmid construct containing virus sequences flanking a short, non-coding linker sequence in place of *lac Z*.
8. Isolation of this recombinant is by colour-based plaque assay and limiting dilution techniques, but additional markers or selection methods will permit more efficient recovery.
9. The gene-deleted virus is characterized by Southern blot analysis. Stocks are prepared by transfection and growth in the complementing cell line, and purified as necessary.

In order to delete the gene from the virus, a plasmid is constructed to carry virus DNA flanking the gene(s) to be deleted. A selectable marker gene, such as *lac Z*, under the control of a eukaryotic promoter, is inserted in place of the deleted gene(s), and the resulting plasmid is co-transfected with purified virus DNA into appropriate cells and recombinants selected. In deriving the gH^- HSV, the thymidine kinase (TK) gene, which is located adjacent to the gH gene, was utilized as a marker, and recombinants with a TK^- (acyclovir-resistant, blue-plaque) phenotype selected (15). This selection for TK^- virus can increase the percentage of recombinant plaques seventy-fold. In a second stage recombination procedure designed to remove the *lac* Z marker gene, purified, defective virus DNA carrying the *lac Z* gene can be transfected into cells along with a plasmid designed for recombination with this virus DNA, allowing removal of the *lac Z* gene. For this step, advantage was again taken of the proximity of the TK gene to the gH gene, by constructing the plasmid so that recombination restored TK activity, permitting selection using methotrexate, and picking white, TK^+ recombinants using a TK^-, gH^+ cell line (15). Following plaque purification and limiting dilution cloning, the replication-defective virus can be characterized by Southern blot analysis, and stocks for use as vaccine prepared to high titre in the complementing cell line and purified as required.

The attraction of replication-defective viruses as vaccines lies in their live virus attributes, particularly intracellular delivery in a form likely to promote Class I and Class II cellular immune responses, especially the generation of cytotoxic T lymphocytes through appropriate virus antigen presentation; the capacity for almost complete virus gene expression; the release from infected cells of intact, although non-infectious, virus particles, and the potential for induction of a broad range of immune responses. Studies in animals with a replication-deficient herpes simplex virus have indicated that delivery to one mucosal surface can induce protection against clinical disease at another (16). The advantages of this virus vaccination strategy are reinforced, compared with live, attenuated, replication-competent viruses, by marked reduction in the possibility of reversion to virulence, by reduced persistence in the host, and by limited exposure of the host to virus DNA. In comparison to subunit virus vaccines, there is no requirement for any adjuvant, carrier, or delivery system.

Replication-defective HSV virus vaccines are currently in clinical trials to determine if the expectations indicated by the preclinical evaluations can be fulfilled in humans. In addition, strains of simian immunodeficiency virus (SIV) lacking sequences coding for the *nef* and other genes, and described as highly attenuated when evaluated in rhesus macaques (17), are under consideration for use as vaccines against HIV-1.

7. Enhancement strategies for virus vaccines

It is generally accepted that live virus vaccines, whether attenuated, reassorted, or replication-deficient viruses, are capable of eliciting a greater, more durable, and probably more relevant immune response than non-living virus vaccine preparations, although other factors, particularly the route of administration, may play an important role. This perception of live virus vaccines as superior immunizing agents compared to non-living preparations is somewhat offset by the associated disadvantages of the former, such as presence of nucleic acid, the possibilities of reversion to virulence, recombination or integration events, and concerns regarding the use of these preparations in immunocompromised individuals. However, the perceived divide between living and non-living vaccines is narrowing, primarily because of two factors: the increasing capability to eliminate unwanted genes and their potentially deleterious effects from live, candidate vaccine viruses by genetic and molecular engineering technology on the one hand, and rational and enlightened use of some form of adjuvant, vector, or delivery system for non-living virus vaccine preparations on the other.

There is no doubt that virus subunits, virus proteins, or other subvirus structures small in size and of low molecular weight relative to an intact virion, whether live or inactivated, are intrinsically poor immunogens. The antigenicity of such materials is also not helped by their relatively pure, lipid-free nature, and their existence in monomeric form. However, the antigenicity of these materials can often be enhanced under experimental conditions by incorporation with some form of adjuvant or carrier. *Table 2* lists several such materials, generically known as adjuvants, delivery systems/carriers, or immunomodulators. Due to the great diversity in the nature of these substances, their grouping may be structured in various ways, and the difficulty in their classification is compounded by their varied mechanisms of action. The constraints and aims of this volume do not permit detailed discussion of this topic; however, there are reviews to which the reader is referred for further information (18,19). What is pertinent is the increasing expectation that, given a virus antigen or antigenic epitope of known molecular structure, physical properties, and immunogenic potential, together with a preferred route of administration, it is becoming increasingly possible to make a rational choice of enhancement strategy for that particular antigen.

Table 2. Agents for enhancement of host immune responses to virus vaccines

Adjuvants	Delivery Systems and Carriers	Immunomodulators
Aluminium salts	Mineral oil emulsions	Cytokines
Bacterial products	Living virus or bacterial vectors	Hormones
Surface active agents	Immunostimulating complexes	Vitamins
Polyanions and polyacrylics	Biodegradable microparticles	Complement fragments
Synthetic antigen constructs	Squalene emulsions	Monoclonal antibodies
	Bacterial toxins and toxoids	
	Liposomes	
	Ty virus-like particles	

Considerations that will influence a rational choice of vaccine antigen-enhancement strategy (*Protocol 7*) will include factors concerned with the nature of the antigen, the ease, convenience and standardization of the technology involved in the interaction of the antigen with the adjuvant or carrier, as well as factors impinging on the activity of the antigen-adjuvant or carrier formulation in the host. Amongst the latter, the route of vaccine administration required, and the local or systemic nature of the virus infection, the required organ(s) targeted by a given virus infection, the immune response most relevant to preventing or eliminating the infection, and any potential immunogenic effects of the adjuvant or carrier itself, must all be taken into account.

Protocol 7. Rational selection of enhancement strategies for virus vaccines

1. The need for an enhancement strategy for a virus vaccine will depend on vaccine antigen nature and size, route of immunization to be used, possibility of degradation by host enzymes, and targeting requirements.
2. The type of virus vaccine enhancers available include promoters of the immune response (adjuvants), non-living delivery systems (designed to augment the immune response through optimal vaccine antigen presentation), living vectors (viruses and bacteria), and immunomodulators such as cytokines.
3. The mechanism of action of a virus vaccine enhancer may be through immunomodulation of host defences, depot formation and prolonged release of antigen, intracellular pathway delivery, local recruitment of cellular defences, cell surface interaction and modification, physical form of antigen presentation, or increased antigen uptake efficiency.
4. Virus vaccine enhancers must be safe and induce no local or chronic inflammation, induce no systemic symptoms or pyrogenicity, be inert and non-immunogenic, be biodegradable, have no carcinogenic or teratogenic effects, and have no capacity for dissemination or persistence in the host or the environment.
5. Problems associated with the use of virus vaccine enhancers include characteristics of the antigen-enhancer interactions, the relevance of the host response, adjuvanted vaccine antigen assay and standardization, lack of animal models for virus vaccine enhancer efficacy that can translate to man, and epitope-specific suppression.

Since the initial encounter of many viruses with the host is at a mucosal surface, the idea of local, mucosal surface delivery of virus vaccines or vaccine antigens, particularly via the oral route, with its advantages of ease of administration and greater acceptability, is an attractive one. Problems with this strategy that need to be overcome are concerned with quantities of vaccine antigen needed, denaturation and proteolytic destruction of antigen, and the development of unwanted immunological effects, particularly antigen tolerance. In spite of these difficulties, a number of adjuvants or carriers, including particulate delivery systems such as biodegradable microparticles and immunostimulating complexes (ISCOMs), as well as proteins such as cholera toxin or its B subunit, which may act through targeting vaccine antigens to lymphoid tissues, thereby enhancing antigen uptake, are showing promise in providing protection and delivery of vaccine antigens to appropriate sites on gut and other mucosal surfaces.

A significant advance in immunoenhancement strategy is the recognition that, depending on the mode of presentation of the virus vaccine antigen to the immune system in terms of its form and amount (and also the local environment existing at the site of antigen deposition), the immune response may be routed or steered in various directions (20, 21). This routing of the immune response, possibly effected at first contact of the host with the vaccine antigen, can elicit a response that may or may not include a cytotoxic T-cell element. Some adjuvants or carrier systems, notably aluminium salts, the only adjuvant currently licensed for use in man, steer the immune response to the associated antigen, at least in mice, towards a Th2-like, humoral immune response, manifested by induction of cytokines such as interleukin-4 and interleukin-10, and subclass IgG_1 antibodies. Other adjuvant formulations, including certain particulate antigen carriers such as ISCOMs, can induce a Th1-like immune response in mice, with production of interferon-gamma (IFN-γ) and interleukin-12 (IL-12), a $CD8^+$ MHC class I-restricted CTL response, as well as relatively high IgG_{2a} subclass antibody responses. This latter is also the type of response believed to be promoted by a live virus vector, into which genes coding for putative protective antigens from other viruses have been engineered, thereby engendering induction of CTL responses to these antigens.

A further enhancement strategy for virus vaccines currently being pursued is the use of natural immunomodulators such as cytokines. Cytokines, injected together with virus antigens as a conjugated or unconjugated preparation, either alone, in some form of particulate carrier such as liposomes or biodegradable microparticles or in live carriers such as vaccinia virus, may steer the immune system towards a Th-1 or Th-2 like response as required, or may provide a prolonged stimulus to the immune response through slow release of antigen. Cytokines such as IFN-γ, interleukin-1 (IL-1), interleukin-2 (IL-2), interleukin-12 (IL-12), and, more recently, interleukin-6 (IL-6), are all under investigation for use as immunoenhancers for virus antigens, and all are reasonably well tolerated in low to medium doses. Problems associated with the use of cytokines as immunoenhancers include their short half-life *in vivo*, and the difficulty in prolonging their desirable effects, although recent studies have shown that injection of IL-12 DNA into mice results in long-lasting expression of this cytokine in serum (22).

8. *In vitro* and *in vivo* models in vaccine development

Vaccine development requires that putative vaccines, whether inactivated, attenuated, or genetically modified, be tested for safety and efficacy before general release. For inactivated vaccines, the requirement is firstly for safety, and this precedes any studies of efficacy. History shows this to be important, since some batches of the earlier poliomyelitis virus vaccine were insufficiently inactivated and induced disease in some recipients, while other vaccines, given to volunteers, were later shown to be contaminated with adventitious agents. Once purity is established, inactivated vaccines can be evaluated *in vitro*, in animal models (23), or in the absence of such models, the target immunization group, for reactions, immunogenicity, and protection against challenge virus infection.

Preclinical evaluation in animal models has been used to test the efficacy of certain inactivated vaccines, including hepatitis B virus vaccines in chimpanzees, influenza virus vaccines in ferrets and mice, hepatitis A vaccine in marmosets and tamarins, and rabies virus vaccines in dogs and foxes. In each case the test species is known to be susceptible to homologous, virulent virus infection, which can be monitored either experimentally, or naturally by exposing animals to an infectious environment. These types of studies suffer from some drawbacks:

- chimpanzees, used to test hepatitis B virus, are not as susceptible to infection as humans
- influenza virus infection in mice either produces no clinical response, or a consolidated pneumonia, both untypical of human disease.

In contrast, the use of dogs and foxes to test inactivated rabies virus vaccine is directly relevant, since both species suffer clinical disease similar to man, and are also a target group for immunization, while ferrets produce the same clinical response to influenza as humans. Some authors have criticized the use of animal models (24), suggesting that the results obtained can be unhelpful and misleading, and that efficacy studies should only be relied upon when carried out in the natural host. Despite this view, the weight of opinion is that preclinical *in vivo* evaluation is helpful in the development of human vaccines, but that the data from such studies cannot be conclusive for the effects in humans; the animal model may be evolutionarily remote from man, and the clinical course of challenge virus infection distinct.

Problems are compounded when studying live attenuated virus vaccines, since the clinical and immunological effects of the vaccine in experimental animals may differ from those in the target immunization species, and the effects of challenge infection may again be distinct. However, some early confidence in the state of virus attenuation can be achieved *in vitro* in a number of ways. Firstly, putative attenuated virus vaccines may possess elements of nucleotide sequence which differ from those of wild-type virus. Thus live, attenuated poliovirus vaccine strains are distinguishable by sequence analysis from wild-type virus, and this principle also applies to both rubella and mumps attenuated virus vaccines, and probably several others. Secondly, most attenuation procedures have been carried out by passage of virus at low temperatures. The ability of the putative attenuated virus to grow at low temperatures, and have no or restricted growth at higher temperatures, is considered an important *in vitro* marker of attenuation, and has been useful for the vaccines of rubella, measles, poliomyelitis, influenza, and dengue 1 and 2 viruses.

Problems are apparent, however, in the development of human vaccines, where animal models are usually sought, and where equating the results of animal studies to human experience may be inappropriate. In some cases no animal model is available, and adenovirus vaccines and the live attenuated virus vaccines against measles, rubella, and varicella were tested directly in phase I studies in human volunteers. Where animal models are known, they have been used. Argentine haemorrhagic fever virus vaccine has been tested in guinea pigs, rhesus and macaque monkeys, and monkeys have also been used to test attenuated poliomyelitis and yellow fever vaccines, where the virus vaccine has been shown to have reduced neurological effects. For influenza vaccines, infection can be demonstrated in monkeys, hamsters, rats, and mice, but no clinical responses occur; on the other hand, ferrets produce a clinical response to live influenza virus similar to that of humans, and for this reason this species is the preferred model for many vaccine developments against this disease.

For animal virus vaccines the natural host represents the optimal system, but information can be obtained in other species. Thus, attenuated borna disease virus vaccine, a virus which causes encephalitis in horses, has been tested in rats, where it induces mild encephalitis. Bovine respiratory syncytial virus vaccine has been tested in cows, and most other animal virus vaccines have also been tested in the natural host. In these cases, the problems posed by using an artificial model do not exist; the vaccines can be shown to induce a significantly reduced clinical effect, and resistance to challenge infection indicates protection against natural disease.

It is recognized that a clinical response to challenge infection is an important factor in the selection of an animal model for vaccine development. Thus, respiratory syncytial virus replicates in monkeys, ferrets, and guinea pigs without any measurable clinical effect, but infection in mice can be shown to alter the pattern of respiratory function, and this response could be used to monitor both the attenuation of vaccine strains and the effect of wild-type virus challenge. Similarly, rabies virus causes clinical illness and death in many animal species which have been valuable in the development of vaccines against this disease. In some cases, the results obtained in animal models can be misleading: the use of monkeys in the development of live poliomyelitis vaccine was limiting, until it was realized that monkeys are more sensitive to poliovirus infection than humans, and the reduced neurological effects of attenuated vaccines in monkeys could be equated with no neurological effects in man. Again, attenuated HSV type 1 virus vaccine in mice confers protection against a virulent, lethal challenge given intraperitoneally, and studies in guinea pigs using a genital route of challenge virus infection reproduces this data. However, in spite of the success of HSV vaccines in experimental laboratory animals, no HSV vaccine has yet proved effective in man.

In vivo evaluation of virus vaccines permits other studies and reveals other problems. The use of ferrets in the development of some putative live attenuated influenza virus vaccines has revealed genetic instability in some of these viruses, with loss of temperature sensitivity, and reversion to virulence. This has led to the rejection of these vaccines. Animal models also allow studies of immunization strategies, vaccine formulations, toxicity, and dosage levels, and putative vaccines can be administered by various routes; all these studies can be accompanied by measurements for safety and immune responses. The use of animal models in vaccine development and evaluation is both widespread and controversial, but for many vaccines has provided valuable data, although this does not necessarily indicate value in humans. Animals have distinct features in their immune systems, their responses may differ from those of man, and they may respond differently to challenge virus infection, in some cases being less sensitive and in others more sensitive than humans. The conclusion is that despite their value, animal models can never replace volunteer studies of vaccine safety or efficacy.

9. Control and licensing of virus vaccines

The regulatory bodies that grant licenses for biological products, including virus vaccines, hold responsibility for ensuring that vaccines are pure, safe, and potent before allowing use and/or sale, and this demands compliance with stringent requirements (25, 26). These regulations have been formulated over a number of years; they are constantly evolving as new forms of products emerge, and further experience of past products accumulates. The need for regulation was first realized in 1901, when a child in the USA died of tetanus contracted from contaminated diphtheria antitoxin. More recently, regulations have been extended to ensure that infections resulting from the inoculation of subjects with inadequately inactivated vaccine, as occurred with two lots of poliovirus vaccine manufactured in 1955 by the Cutter Laboratories in California, should never occur again (27). In addition to the evolution of license requirements from past experience, judgement by licensing authorities is paramount, since no two vaccine preparations are exactly the same, and new preparations constantly pose new problems. The responsibility for regulation is borne by National Government Agencies, such as the Food and Drug Administration in America, and similar authorities in most developed countries. The World Health Organization Standards Committee sets criteria which are accepted by some hundred individual countries, and the European Community will probably set up a common licensing authority in due course. Although each authority is distinct, the requirement for purity, safety, and potency is common; thus, regulations tend to be similar.

The process of developing a virus vaccine from the point of recognition of need to the point of licensing can be considered in a series of stages. Initially, there is the development of the vaccine. As outlined above, this stems from knowledge of the commonness and severity of the infection. For example, human herpes viruses, mumps, and measles affect the majority of people in all countries, and a significant proportion of these are clinical infections: influenza can infect up to 70% of a population in epidemics, and cause over a million deaths during pandemics; rubella infection during pregnancy is devastating; poliomyelitis in the first half of the twentieth century caused annual epidemics with high mortality or paralysis; and yellow fever is a life-threatening infection of equatorial countries. The method of vaccine preparation is individual, and vaccines may be non-living, such as influenza, live attenuated preparations such as measles and mumps, or genetically engineered such as foot-and-mouth disease and hepatitis B vaccines. The development includes studies of inactivation, stability, laboratory markers, and animal tests, and these result in a high degree of confidence in the safety and efficacy of the product. Thus, attenuated vaccines can be identified in the laboratory for markers such as temperature sensitivity, and in animals to test for toxicity, stability, immune response, and protection against challenge virus infection. Animal models can also be used to test non-living vaccines for the same properties. At this point, it is now appropriate to consider volunteer studies.

Virus vaccines for use in humans require license from regulatory bodies, and can only be prepared in prescribed facilities which are both licensed and monitored by the regulatory authorities. The buildings and facilities where vaccines are to be prepared must conform to specific design and construction requirements. The rooms used for storage, labelling, manufacture, and testing require approval, as do all areas in which operations are carried out. Sterile operations and facilities need to be approved, together with the flow of personnel working in these areas; water and air supplies require validation. Where animals are used in the production or evaluation of vaccines, quarantine regulations are required, together with adequate methods for disposing of animal waste. Controls must be built in and maintained for all viruses brought into the area, and include procedures to ensure against contamination.

All equipment, including calibration equipment for autoclaves, dry heat ovens, freezers, and incubators must be approved. Adequate procedures to prevent cross-contamination must be in place. All materials must be clearly labelled, together with methods to insure against human error. There must be approved storage facilities, with a system for retention of samples. Lastly, complete records of all procedures must be maintained. These regulations are strict: commercial companies build facilities specifically to meet the requirements, with advice from the licensing authorities, and the continuous monitoring of standards as an essential part of the usage. Given that the standards are met and maintained, license to use the premises for the production of vaccine for human use will be approved, and vaccine manufacture can proceed. The reader is referred to review articles for further information (25, 26).

Although the regulations that govern the licensing of premises for vaccine production are similar for different authorities, those that control the production of vaccines vary considerably; some are generally applicable, but others are specific to particular vaccines. For non-living virus vaccines, the kinetics of inactivation must be known. This is important, and stems from the Cutter incident. Inactivation must be demonstrated for production batches, and in all production methods; scaling up from the laboratory bench must be shown to be successful, and not alter vaccine properties and characteristics. Vaccine must be purified, and this must be proved by biochemical assays such as gel electrophoresis, peptide mapping, chromatography, or immunological tests such as Western blotting, or a combination of such tests. The vaccine must be shown to be stable. Samples are taken at intervals, and shown to be sterile by prolonged culture in thioglycollate broth, and free of mycoplasmas. The vaccine must be inoculated into animals as a measure of general safety, and the Limulus amoebocyte test is used to ensure freedom from endotoxin and pyrogens. Finally, the vaccine must be shown to be potent in mouse tests. For production of DNA vaccines, and certain genetically engineered vaccines, safety tests will also include evidence that the nucleic acid is not incorporated into host chromosomes, and that there is no evidence of autoimmune responses following immunization.

For live, attenuated virus vaccines, the virus must be grown on an approved cell substrate and repeatedly cloned for purity; the cloned vaccine must itself be shown to be free of adventitious agents and mycoplasmas. Vaccine must be relatively thermostable and genetically stable. These factors are commonly determined in earlier laboratory studies, but must be repeated with bulk-grown virus. Precautions include the inoculation of animals to ensure safety, attenuation, and toxicity, and the vaccine must be shown to have retained *in vitro* markers of attenuation, if these exist. Finally, the vaccine must be cultured in cells and shown to be neutralized by a specific antibody; no virus should be recoverable from the neutralized material, and this material must be sterile.

Candidate vaccines for use in humans, which have satisfied the above tests, and have the approval of licensing authorities, are now subject to three stages of clinical studies, phases I, II, and III, which must be carried out sequentially and successfully before general licence is granted. Phase IV clinical studies proceed after license is granted. Phase I studies measure safety and immunogenicity: the vaccine is given to a small number of volunteers from the target population, and shown not to induce unacceptable reactions and to induce antibody and, increasingly, other immune responses such as evidence of CTL activity, considered of importance in protection, as measured by *in vitro* tests. Satisfactory conclusion of these tests allow phase II studies to proceed, where the optimal dose for an effective immune response is calculated, using larger numbers of volunteers. Phase III studies may involve several hundreds or thousands of participants; these studies again measure safety and immunogenicity for the optimal dose of vaccine, and evidence of protection can be acquired by demonstrating that immunized people are protected against disease following natural exposure. Studies in specific target or high-risk groups may be undertaken as part of the phase II or phase III clinical trials. The successful completion of phase I to phase III studies can give the vaccine license for general use, but this is not the final assessment. Licensing requires that general use is carefully monitored for some years for possible reactions that may occur in certain groups with different ethnic or medical backgrounds, and the duration of immunity is followed: these are phase IV studies. The long-term follow-up is important. Subsequent to the early introduction of an inactivated measles virus vaccine, which in short-term clinical trials gave protection, evidence accumulated to indicate that a modified form of measles could occur in later life; and the finding of encephalitis in patients given one form of live mumps virus vaccine was only found after general licence was authorized, and led to the withdrawal of this vaccine.

Protocol 8. Requirements for licensing of virus vaccines for general use

1. Seed virus developed through one of the above methodologies, grown in a licensed substrate, plaque-purified and free of adventitious agents, is made available.
2. Transfer of seed virus to licensed premises.
3. For non-living vaccines:
   (a) Produce bulk viruses from seed virus in licensed substrate.
   (b) Virus is inactivated.
   (c) Vaccine virus is purified; and demonstrated to be pure using one or more biochemical tests.
   (d) Inoculate into animals to ensure safety and demonstrate antigenicity.
   (e) Limulus amoebocyte test to ensure vaccines are free of endotoxin.

For live vaccines:

(a) Produce bulk virus from seed virus in licensed substrate.

(b) Ensure bulk virus is free of contaminating, adventitious agents, and mycoplasmas.

(c) Animals are inoculated to ensure attenuation.

(d) *In vitro* tests to ensure virus is completely neutralized by antiserum.

Each of the above steps is accompanied by sterility tests, documentation, retention and storage of samples.

4. Clinical studies on human vaccines: Phase I – small numbers (about 10–20) of volunteers inoculated with vaccine to detect and measure unacceptable reactions and immunogenicity; test for possible reversion to virulence of live attenuated vaccines.
5. Phase II clinical studies: larger numbers of volunteers (100–200) inoculated to evaluate optimal dose of vaccine and immune responses.
6. Phase III clinical studies: large number of volunteers inoculated with standard dose of vaccine and monitored for immune responses, and, if possible, protection against homologous virus infection.
7. Phase IV clinical studies: monitor for unexpected reaction and immunity for some years after general licence has been granted.

10. Future Developments

The future for immunization against virus diseases can be considered under three headings: improvements to existing vaccines through the application both of novel strategies of immunization and of innovations in molecular and genetic engineering; development and implementation of global strategies for vaccination, such as eradication policies, the WHO Global Programme for Vaccines and Immunization, the Children's Vaccine Initiative (CVI), and vaccine combination strategies; introduction of novel concepts which may have application to a wide range of vaccines, such as 'skin immunization', the activation and amplification of immune cell interactions through oxidative mitogenesis, and oral immunization with genetically modified plant foods.

Influenza virus vaccines represent one of the least effective vaccines currently licensed for use in man, and considerable efforts to improve these vaccines are in progress. One strategy for this, and indeed for other existing or newly developed vaccines, is the inclusion of new adjuvants or immunomodulators such as QS21 (derived from saponin), monophosphoryl lipid A, or MF59 (a squalene oil and water emulsion plus a detergent), with the aim of increasing and broadening immunogenicity, as well as targeting the antigens to stimulate cellular and/or humoral immune responses. This approach may also be especially effective when certain cytokines are incorporated with virus antigen(s) in a particulate structure such as virosomes or ISCOMs. Other strategies for influenza virus vaccines, in particular, include preparation of a vaccine grown in cell culture rather than embryonated eggs. This bypasses the problems of egg allergy and vaccine availability during pandemics or widespread epidemics, providing a more antigenically relevant vaccine and increasing vaccine dosage. Finally, the nasal or other mucosal route may be used for vaccine administration, with virus antigens packaged in some form of carrier providing protection, permitting increased adherence or uptake, and acting to enhance antigenicity.

Global strategies for immunization against virus diseases have already proved successful with the eradication of smallpox, and similar global eradication programmes are currently in operation for measles and poliomyelitis, with the aim of eliminating these infections world-wide in the early twenty-first century. The CVI launched in the 1980s was aimed at facilitating the development and use of vaccines, including virus vaccines, in childhood immunization programmes world-wide (28). However, economic considerations have somewhat blurred the impact of the CVI, and in attempts to resolve these problems much recent work has focused on the administration of vaccines combined into a single formulation. Although a very worthwhile goal, already in sight through the current use of a number of combined vaccine preparations for children including the diphtheria, pertussis, and tetanus (DPT) vaccine against bacterial diseases, the triple poliomyelitis vaccine, and the measles, mumps, and rubella (MMR) vaccine against virus diseases, the expansion of this strategy to encompass other more recently developed bacterial (Haemophilus influenza type b, and acellular pertussis) and virus (hepatitis type B and hepatitis type A) vaccines has encountered problems, probably immunologically based, but manifested as interference between vaccine preparations. Research into understanding and hopefully resolving these problems is already underway.

Novel concepts for immunization against infectious diseases include the enhancement of the immune response to antigens through the use of oxidative mitogenesis (29) – the generation of reactive aldehydes at the T-lymphocyte cell surface, and the consequent increased level of activation of these cells. Schiff's base-forming molecules such as tucaresol, a small xenobiotic, substituted benzaldehyde, can act at the T-cell surface, allowing amplification of normal cell physiological processes, and providing immunopotentiation through increasing reactogenicity of the T-lymphocytes with antigen presenting cells. Skin immunization using cholera toxin as a transcutaneous adjuvant has recently been reported as a means of increasing the ease of administration of vaccines and avoiding some of the logistic problems associated with vaccine delivery through parenteral inoculation. In preliminary studies in mice, it has been reported that cholera toxin acts in its adjuvant capacity, as well as facilitating passage across the skin and stimulation of an immune response to a co-administered antigen, by merely wetting the skin surface with the immunizing solution (30).

In recent years the application of genetic engineering to the development of transgenic plants expressing virus (and other microbial) proteins such as the hepatitis B surface antigen, the rabies virus glycoprotein, and Norwalk virus capsid protein, has opened up the possibilities of oral immunization against virus diseases through genetically engineered foodstuffs as edible vaccines (31). Although several problems surrounding this strategy remain to be addressed, the concept is an attractive one for overcoming the difficulties of mass immunization in developing countries.

These, plus other novel concepts of immunization, in association with the continued application of advances in molecular biology and genetic engineering to the development of novel virus vaccines, and the increasing application of global immunization strategies, may result in further successes in the control of both existing and newly emerging virus diseases in the twenty-first century.

References

1. Plotkin, S. A., and Mortimer, E. A. Jr. (1994). *Vaccines* (2nd edn). W. B. Saunders Company, Philadelphia.
2. Stephenne, J. (1990). In *Virus Vaccines* (ed. A. Mizrahi), p. 279. Wiley-Liss, NY.
3. Delves, P. J., Lund, T., and Roitt, I. M. (1997). *Mol. Medicine Today*, **3,** 55.
4. Potter, C. W. (1994). *Rev. Med. Virol.,* **4,** 279.
5. Palese, P., Zavala, F., Muster, T., Nussenzweig, R. S., and Garcia-Sastre, A. (1997). *J. Infect. Dis.*, **176 (S),** S45.
6. Joensuu, J., Koskenniemmi, E., Pang, X-L., and Vesikari, T. (1997). *Lancet*, **350,** 1205.
7. Perkus, M. E. and Paoletti, R. (1996). In *Concepts in Vaccine Development* (ed. S. H. E. Kaufmann), p. 379. Walter de Gruyter, Berlin.
8. Mackett, M. (1990). *Seminars Virol.,* **1,** 39.
9. Heineman, T. C., Connelly, B. L., Bourne, N., Stanberry, L. R., and Cohen, J. (1995). *J. Virol.,* **69,** 8109.
10. Hanke, T., Blanchard, T. J., Schneider, J., Hannan, C. M., Becker, M., Gilbert, S. C., Hill, A. V. S., Smith, G. L., and McMichael, A. (1998). *Vaccine*, **16,** 439.
11. Manickan, E., Karem, K. L., and Rouse, B. T. (1997). *Crit. Rev. Immunol.,* **17,** 139.
12. Donnelly, J. J., Ulmer, J.B., Shiver, J. W., and Liu, M. A. (1997). *Annu. Rev. Immunol.,* **15,** 617.
13. Robertson, J. S. (1994). *Vaccine,* **12,** 1526.
14. Forrester, A. J., Farrell, H., Wilkinson, G., Kaye, J., Davis-Poynter, N., and Minson, A.C. (1992). *J. Virol.,* **66,** 341.
15. Bournsell, M. E. G., Entwisle, C., Blakeley, D., Roberts, C., Duncan, I. A., Chisholm, S. E., Martin, G. M., Jennings, R., Ní Challináin, D., Sobek, I., Inglis, S. C., and McLean, C. S. (1997). *J. Infect. Dis.*, **175,** 16.
16. McLean, C. S., Ní Challináin, D., Duncan, I., Bournsell, M. E. G., Jennings, R., and Inglis, S. C. (1996). *Vaccine*, **14,** 987.

17. Wyand, M. S., Manson, K. H., Garcia-Moll, M., Montefiori, D., and Desroisiers, R. C., (1996). *J. Virol.*, **70,** 3724.
18. Newman, M. J., and Powell, M. F. (1995). In *Vaccine design: the subunit and adjuvant approach* (ed. M. F. Powell and M. J. Newman), p. 1. Plenum Press, NY.
19. Jennings, R., Simms, J. R., and Heath, A. W. (1998). In *Modulation of the immune response to vaccine antigens* (ed. F. Brown and L. R. Haaheim), p. 19. Karger, Basel.
20. Ahlers, J. D., Dunlop, N., Alling, D. W., Nara, P. L., and Berkofsky, J. A. (1997). *J. Immunol.*, **158,** 3947.
21. Baldridge, J. R., and Ward, J. R. (1997). *Vaccine*, **15,** 395.
22. Moelling, K. (1998). *Gene Therapy,* **5,** 573.
23. Mäkelä, H., Mons, B., Roumiantzeff, M., and Wigzell, H. (1996). *Vaccine*, **14,** 717.
24. Hilleman, M. R. (1998). *Vaccine*, **16,** 778.
25. Davenport, L. W. (1995). In *Vaccine design: the subunit and adjuvant approach* (ed. M. F. Powell and M. J. Newman), p. 81. Plenum Press, NY.
26. Parkman, P. D., and Hardegree, M. C. (1994). In *Vaccines* (ed. S. A. Plotkin and E. A. Mortimer, Jr.), p. 889. W. B. Saunders, Philadelphia.
27. Nathanson, N., and Langmuir, A.D. (1995). *Am. J. Epidemiol.,* **142,** 109.
28. Shepard, D. S., Walsh, J. A., Kleinau, E., Stansfield, S., and Bhalotra, S. (1995). *Vaccine*, **13,** 707.
29. Chen, H., and Rhodes, J. (1996). *J. Mol. Med.,* **74,** 497.
30. Glenn, G. M., Rao, M., Matyas, G. R., and Alving, C. R. (1998). *Nature*, **391,** 851
31. Arntzen, C. J. (1998). *Nature Med.,* **4,** 502.

7

Antiserum production and monoclonal antibodies

J. MCKEATING, C. SHOTTON and M. VALERI

1. Introduction

Since 1975, when Kohler and Milstein first reported the immortalization of antibody-secreting cells by somatic cell hybridization, hybridoma technology has proved to be an invaluable tool in the isolation and characterization of a variety of intra- and extracellular antigens (1). The principle is to propagate a clone of cells from a single antibody-secreting B lymphocyte, so that a homogeneous preparation of antibody can be obtained in large quantities. Hybrid cells are formed by the fusion of B lymphocytes (isolated from the lymph nodes or spleen of immunized animals) with a B-cell tumour line, most commonly, from the same species. The resultant hybridoma, once selected for specificity, will secrete the required antibody and can be grown indefinitely. Monoclonal antibodies (mAbs) produced against specific virus antigens have been used extensively to define the structure, function of virus components, and their interaction with proteins of virus and host-cell origin. Furthermore, such mAbs can form the basis of assays for the diagnosis of virus infection.

Viruses generally have a regular, repeated, surface structure, and their surface-expressed antigens generally have a critical role in the attachment of virus to the host cell. Hence, such antigens provide a prime target for the immune system, and antibodies produced in response to these antigens have the potential to neutralize virus infectivity. These antibody–antigen interactions help define regions of the virus responsible for cell attachment and entry.

In this chapter we will discuss different immunization regimes, and the production and purification of both polyclonal and mAbs to virus antigens.

2. Antigen preparation

Virus proteins from a variety of sources may be used for immunization studies. However, purified protein is more likely to elicit the desired antibody response, certainly in cases where poorly immunogenic molecules are being studied (Chapter 3). It is critical to ensure that the immunogen is as closely related to the 'native' protein as possible, especially if the antibodies will be required to interact with intact virus particles or infected cells. At the start of a new immunization regime, the investigator needs to consider the use(s) of the antibodies to be generated, to ensure the optimal choice of immunogen, and subsequent screening assays for mAb selection. Immunogens can include: synthetic peptides, recombinant virus proteins expressed in prokaryotic or eukaryotic systems, purified virus proteins, or intact whole cells. If antibodies are required against virus proteins which undergo extensive post-translational modification(s), such as glycosylation or myristoylation, this should be taken into account when choosing expression systems for the generation of recombinant proteins. If, for example, one wanted to raise a neutralising antibody response, specific for the human immunodeficiency virus (HIV) glycoprotein gp120-gp41, one may choose a recombinant protein synthesized in mammalian cells to mimic the native glycosylated protein, or whole cells engineered to express the glycoprotein at the cell surface. Peptides or prokaryotic expressed proteins (non-glycosylated) would be of little use to generate antibodies capable of recognising native gp120. Alternatively, if one wanted to raise an antibody to a non-structural virus protein for intracellular confocal microscopic studies, a peptide immunogen may be ideal. Hence consideration of the native structure of the protein one wishes to immunize with is important to ensure the successful generation of 'useful' antibodies. Antibodies to both linear and conformation-dependent epitopes are more likely to result from the immunization of correctly folded proteins, protein fragments, or intact cells, and may mimic the humoral response seen in a natural infection.

2.1 Synthetic peptides

Synthetic peptides specifically manufactured to a desired sequence are commercially available. Peptides of 20–30 amino acids or shorter are likely to be poor immunogens, since they may lack sequences recognized by T-helper cells which are essential for generating an antibody response. For this reason it is usual to conjugate small peptides to a larger carrier protein, usually a hapten, which contains many T-cell reactive epitopes, e.g. keyhole limpet haemocyanin (KLH) or ovalbumin (OVA). Glutaraldehyde is used to conjugate the peptides and hapten, either randomly, or by disulfide linkage(s) when the peptide has been synthesized with a terminal cysteine. In the latter case, the peptide will be attached via one end of the carrier protein, and this orientation may restrict immunogenicity. The conjugated peptides are prepared with adjuvant, and immunized by the normal protocols (see section 4.1). In our experience, the use of conjugated peptides often results in an antibody response which only recognizes the immunising peptide, and the antibodies are unable to bind to the native virus protein. It is worth noting that smaller rodents do not seem to respond as well to peptide immunizations as rabbits.

Protocol 1. Peptide conjugation

Equipment and reagents

- Carrier protein at 5 mg ml^{-1} in sterile PBS (e.g. keyhole limpet haemocyanin, ovalbumin, or *Limulus* haemocyanin) (Sigma)
- Chosen peptide at 10 mg ml^{-1} in sterile PBS
- 1 M glycine, pH 6.4
- Glutaraldehyde, 25% aqueous solution (Sigma)
- Sterile phosphate-buffered saline, pH 7.2 (PBS)

Method

1. To 500 μl protein and 200 μl peptide in a glass flat bottomed bijou bottle, add 5 μl glutaraldehyde. (N.B. Glutaraldehyde should be used in a fume cabinet.)
2. Add a small magnetic 'flea', and stir for 15 min at room temperature.
3. Add 100 μl 1 M glycine, pH 6.4, and stir for 15 min.
4. Add 200 μl sterile PBS. Store at −20°C.

3. Production of antisera

Polyclonal antisera are derived from immunized animals, and contain a mixture of antibodies produced against a variety of epitopes present within the immunogen, so that the complexity of the antibody response will be partly determined by the nature of the immunogen. The antibodies produced will vary in specificity, affinity, titre, and isotype, and may react against many discrete epitopes/antigens. The serum is a heterogeneous mix of antibodies produced from multiply-stimulated B-cell clones, and may vary considerably between animals immunized with the same immunogen. Individual sera must therefore be tested for reactivity in the immunoassay for which they are required. Without purification, specific antibodies in the immunoglobulin fraction may only constitute approximately 20% of the total activity. This lack of specificity will reduce efficiency in some assays, and may be responsible for a high background reading in some situations. The most commonly used animals for the production of polyclonal sera are sheep, goats, or rabbits. Rabbits have the advantage of being easy to house and handle.

3.1 Immunization

Experimental use of animals in research is covered by animal welfare legislation. Investigators must ensure that they have met all local and national legal requirements before any immunization protocol is begun. In the case of rabbits, animals between 9 and 10 weeks old, weighing approximately 2.5–3 kg, are used.

Protocol 2. Immunization of rabbits

Equipment and reagents

- Complete Freund's adjuvant (Sigma)

Method

1. Day 1: pre-bleed 15 ml from the ear artery. Retain this sample as pre-immune serum for control purposes.
2. Day 1: immunize with 250 μg antigen plus Complete Freund's adjuvant subcutaneously at four sites, 0.25 ml per site.
3. Day 28: boost with 250 μg antigen plus Complete Freund's adjuvant subcutaneously at four sites, 0.25 ml per site.
4. Day 35: test bleed 15 ml from the ear artery.
5. Day 56: boost as on day 28.
6. Day 65: test bleed 15 ml from the ear artery.
7. Day 84: boost as on day 28.
8. Day 95: test bleed 15 ml from the ear artery.
9. Sampling may be continued as above at regular intervals until day 182, when the animal must be exsanguinated.
10. An optimum of 10 days should be left between the final boost and culling of the rabbit.

The reactivity of rabbit polyclonal serum can be improved by the removal of unwanted contaminating serum proteins. This can be achieved by passing the serum through a protein-G column. Antibodies will bind to the column via their Fc fragment, whilst other extraneous proteins will pass directly through the column. Antibodies can then be eluted by appropriate means. The specificity of polyclonal serum can also be improved by affinity purification, involving linking the immunogen or the protein/peptide of interest to a Sepharose column (see *Protocol 3*). The heat-inactivated serum is then passed through the column, which is then washed to remove unbound proteins before eluting the specific antibodies. This procedure will not result in a clonal antibody preparation, since antibodies to several different epitopes will still be present. However, it is a useful method to improve the specificity of polyclonal serum.

Protocol 3. Preparation of an affinity column

Equipment and reagents

- 500 μg antigen
- CNBr-activated Sepharose 4B (Pharmacia Biotech)
- 1 mM HCl
- 0.1 M sodium acetate, pH 4.0 (coupling buffer)
- 1 M ethanolamine, pH 9.0
- Phosphate-buffered saline, pH 7.5–8.0 (PBS)
- PBS-0.1% sodium azide (PBSA)
- 3 M ammonium thiocyanate

Method

1. Mix 1.5 g CNBr–Sepharose with 200 ml of 1 mM HCl, and incubate for 15 min at room temperature.
2. Collect the slurry on a sintered glass filter, and drain until a moist cake is formed. Add the cake (typically 5 ml volume) to 5 ml PBS containing 500 μg antigen.
3. Agitate gently on a rotating platform for 2 h at room temperature. (Do not use a magnetic stirrer as this generates fines which will slow or block the column.)
4. Centrifuge the slurry at 2000 r.p.m. for 4 min.
5. Wash with 20 ml PBS-0.1 M sodium acetate, pH 4.0.
6. Add 20 ml of 1 M ethanolamine, and agitate for 2 h at room temperature.
7. Spin at 2000 r.p.m. for 4 min. Decant the supernatant.
8. Store as a 50% slurry in PBS-0.1% azide at 4 °C.
9. Heat inactivate the rabbit polyclonal serum at 56°C for 50 min.
10. Load inactivated serum onto the affinity column.
11. Wash column with PBSA until the optical density returns to baseline.
12. Elute antibody with 5 ml of 3 M ammonium thiocyanate.
13. Wash column with PBSA.
14. Collect fractions in approximately 5 ml volumes.
15. Read the optical density of fractions in a spectrophotometer at 280 nm.
16. Dialyse antibody fractions against 2 litres PBSA to remove thiocyanate, changing the buffer daily for 3 days.

4. Production of monoclonal antibodies

The fusion of spleen cells or mesenteric lymph node cells (in the case of Peyer's patch immunization, *Protocol 4*) with neoplastic tumour-derived B cells produces hybrid cells, or hybridomas, which exhibit properties of both parental cell lines. Hybridoma cells have the ability to produce antibodies indefinitely

Table 1. Myeloma and hetero-myeloma cell lines

Name	Species/origin	Reference
Mouse		
P3-X63/Ag	BALB/C mouse	8
NSI/1.Ag 4.	BALB/C mouse	3
X63/Ag 8.65	BALB/C mouse	3
Sp2/0	SP2 mouse hybridoma	3
NS0/1	NSI/1.Ag 4.1	3
Rat		
Y3-Ag 1.2.3.	Lou/wsl rat myeloma	3
IR984F	Lou/wsl rat myeloma	4
Human		
LICR-LON-HMy2	Plasma cell leukaemia	5
Bovine		
	Calf-NS0 heterohybridoma	6
Ovine		
1C6.3a6T.1D7	Sheep-NS0 heterohybridoma	7
Equine		
	Horse-NS0 heterohybridoma	8
Rabbit		
240E-1	Rabbit plasmacytoma 240E-1	9

upon propagation in culture. The cells are fused using polyethylene glycol (PEG 1500) to eliminate the surface tension which normally repels cells, thereby promoting membrane fusion. Since cell fusion is a random process, a means of selecting the desired hybrids is essential. The myeloma cell lines lack the ability to produce the enzyme hypoxanthine–guanine phosphoribosyl transferase (HGPRTase). The selectivity uses the cells' ability to synthesize nucleic acids. This can be achieved by *de novo* synthesis, which is inhibited by aminopterin, or by the backup salvage pathway, which is dependent on the enzyme HGPRTase utilizing hypoxanthine and thymidine. The selection medium containing hypoxanthine, aminopterin, and thymidine (HAT) kills cells lacking the HGPRTase. Unfused lymphocytes die in culture naturally after 1–2 weeks. Therefore only cells resulting from fusion of the lymphocytes, which innately possess HGPRTase, and the immortal myeloma cell line will grow in the selection medium, and some of these cells will have the antibody-producing capacity of the B lymphocytes. Several myeloma cell lines have been developed for use in hybridoma technology, and are listed in *Table 1*. For the purposes of this publication, we will describe the production of rodent hybridomas: mouse, rat, and rabbit.

4.1 Immunization

Rats of any strain aged between 10 and 12 weeks, BALB/c mice aged 6–8 weeks, or rabbits aged 9–10 weeks (2.5–3 kg) are used in all experiments. The route of immunization will depend on the animal and the nature of the immunogen. In our experience, cells from immune lymph nodes are an excellent source of antigen-specific B cells. Cells are routinely taken from the mesenteric nodes of rats immunized via the Peyer's patches, which lie along the small intestine (2). Analysis of test bleeds, usually taken after the second immunization, will give an indication of the level of the immune response elicited, enabling subsequent immunization regimes to be planned (*Figure 1*).

Whole cells are usually injected suspended in phosphate-buffered saline or in growth medium, without the addition of adjuvant. When using soluble proteins a 1:1 mixture with adjuvant is used to give a stable emulsion. The most

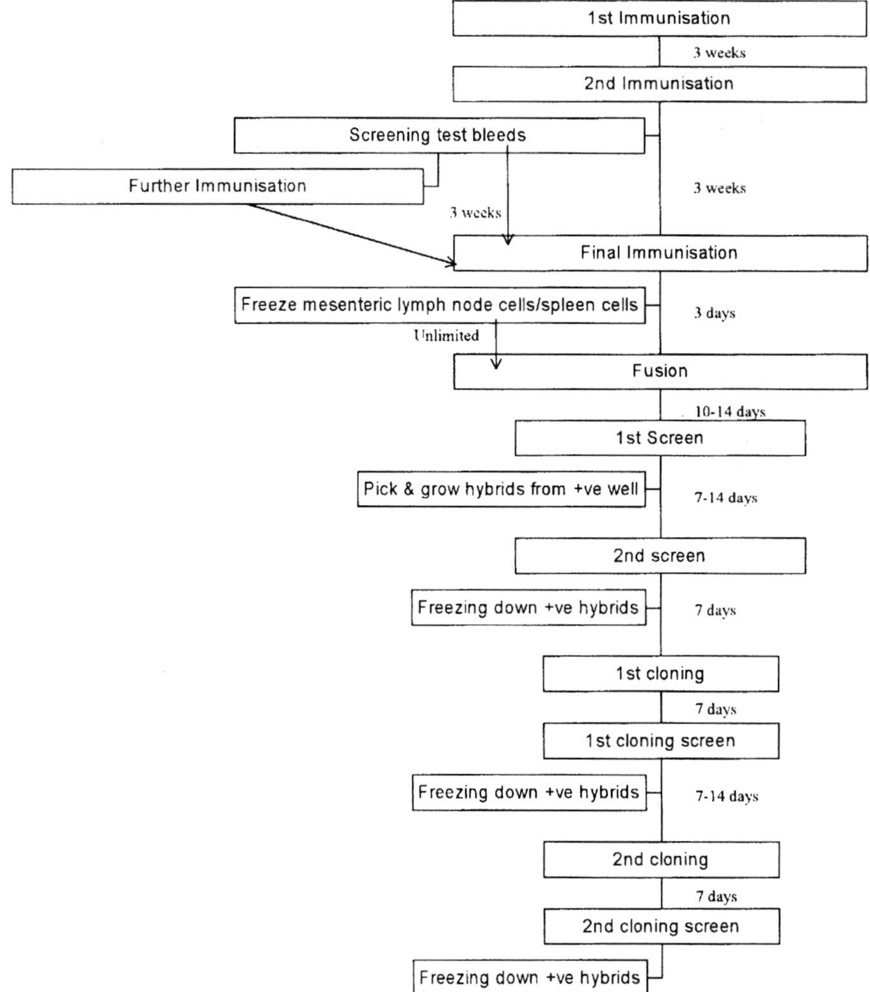

Figure 1. Immunization protocol for mice and rats.

commonly used adjuvant is Freund's, which is an oil (croton oil) containing mycobacteria (Freund's complete adjuvant, FCA), which acts as a slow-release agent, preventing dispersion of the soluble immunogen. It also elicits a strong cellular infiltrate of neutrophils and macrophages at the site of injection. Further immunizations are made using an emulsion with Freund's incomplete adjuvant, which lacks the mycobacteria. Other adjuvants such as aluminium sulfate can be used, and these are less toxic to the recipient animals.

Protocol 4. Immunization for monoclonal antibody production

Equipment and reagents

- Approximately 250 μg antigen
- Complete Freund's adjuvant (Sigma)
- Incomplete Freund's adjuvant (Sigma)

A. *Peyer's patch immunization of rats*

1. Day 1: pre-bleed 1 ml of blood from jugular vein. Retain this sample as preimmune serum for control purposes.
2. Day 1: Peyer's patch immunization: approximately 50 μg antigen plus Complete Freund's adjuvant.
3. Day 21: Peyer's patch immunization: approximately 50 μg antigen plus incomplete Freund's adjuvant.
4. Day 28: test bleed 1 ml of blood from jugular vein.
5. Day 42: intraperitoneal immunization: antigen plus incomplete Freund's (this stage is optional and is dependent on the test bleed result on day 28).
6. Day 49: test bleed.
7. Day 63: Peyer's patch immunization as day 21.
8. Day 66: exsanguinate the animal, and remove the mesenteric lymph nodes.
9. An optimum of 3 days should be left between the final boost and culling of the rat.

B. *Immunization of mice or rats*

1. Day 1: approximately 50 μg antigen emulsified in Complete Freund's adjuvant is injected subcutaneously and intraperitoneally.
2. Day 22: boost as on day 1 with incomplete Freund's adjuvant.
3. Day 43: boost as on day 22 with incomplete Freund's adjuvant.
4. Day 64: 50 μg antigen in PBS is administered intravenously.
5. Day 67: exsanguinate the animal and remove the spleen.

C. *Immunization of rabbits*

1. Day 1: pre-bleed 15 ml from the ear artery. Retain this sample as preimmune serum for control purposes.
2. Day 1: immunize with approximately 500 μg antigen emulsified in Complete Freund's adjuvant injected subcutaneously, intramuscularly, and intraperitoneally.
3. Day 21: boost as on day 1, using incomplete Freund's adjuvant.
4. Day 28: test bleed 1.5 ml from the ear artery.
5. Day 42: boost as on day 21, dependent on test bleed results.
6. Day 56: final boost with 500 μg antigen in saline, administered intraperitoneally and intravenously.
7. Day 60: exsanguinate the animal and remove the spleen.

Fusion efficiency can be increased by incubation of the isolated spleen lymphocytes with the immunising antigen in the presence of CD40L-transfected cells (1/10 the number of spleen cells; X-irradiated) for 24–48 h on plastic dishes to give an *in vitro* boost.

Adherent cells can grow excessively from rabbit spleen, making it difficult for hybridomas to become established (9). The growth of these cells can be limited if the lymphocytes are gently teased from the spleen, by allowing the spleen to swell by injecting PBS through a fine needle and then squeezing the cells out gently. It is not recommended to crush the tissue between slides, nor to force it through a sieve. Alternatively, the excessive growth of adherent cells can be minimized by prior incubation of the spleen cells for 24–48 h on plastic dishes. Non-adherent cells can then be isolated. Should adherent cells continue to interfere with the growth of hybridomas, it may be necessary to transfer the hybridomas to fresh tissue culture dishes. Alternatively, to avoid the problem of adherent spleen cells, rabbits can be immunized subcutaneously in the leg below the popliteal nodes, which are then taken for fusion.

Protocol 5. Immunization of rabbit popliteal lymph nodes

Equipment and reagents

- Complete Freund's adjuvant (Sigma)

Method

1. Day 1: pre-bleed 15 ml from the ear artery. Retain this sample as preimmune serum for control purposes.
2. Day 1: inject 500 μg antigen plus Complete Freund's adjuvant subcutaneously in several sites of the leg below the popliteal nodes.

Protocol 5. *Continued*

3. Day 6: inject 500 μg antigen in phosphate-buffered saline subcutaneously in several sites of the leg below the popliteal nodes.
4. Day 10: test bleed 15 ml from ear artery.
5. Day 11: inject 500 μg antigen in phosphate-buffered saline as on day 6.
6. Day 16: inject 500 μg antigen in phosphate-buffered saline as on day 6.
7. Day 20: exsanguinate the animal and remove the popliteal nodes.

4.2 Analysis of immune response

Periodic test bleeds are usually taken from the tail vein of immunized mice or rats, or from the ear artery of rabbits, to be tested for immune recognition of the virus antigen under test. At this stage the sera can be screened by ELISA, or a suitable biological assay, for reactivity against the immunising antigen and appropriate positive and negative control antigens, if available. It is important to consider protein conformation when establishing such screening assays; often, proteins bound directly to EIA plates adopt random conformation(s), which may differ from the native state presented to the animal during immunization. It is advisable to establish a capture EIA, where an antibody or lectin (for glycoproteins) can bind the antigen to the plastic in a uniform non-denatured manner.

The first sample bleed should be taken one week after the second immunization. The specific antibody titre should be compared with the pre-immune sera from the same animal. A non-specific response may sometimes be observed, resulting in an increased binding of the immune sera to many antigens. On the basis of these results, further immunizations can be planned. A high-specificity antibody titre at this stage signifies that a single additional immunization should be given, and should be sufficient to enable the production of specific monoclonal antibodies. A weak or absent response suggests that at least two more immunizations are necessary. In the UK, Home Office animal welfare regulations limit the total number of immunizations to four. A second test bleed should therefore be taken one week after the third immunization, and compared with the first test bleed and the pre-immune sera. A significant increase in titre at this stage is optimal. The animal should be exsanguinated, and the lymph nodes or spleen removed, three days after the fourth and final immunization. This final test bleed is a useful source of polyclonal sera, and can give an indication of the strength of the immune response prior to the generation of monoclonal antibodies. It is seldom worthwhile to fuse antibody-secreting cells from animals which do not display a detectable serum-specific antibody response. However, the presence of a polyclonal antibody response does not guarantee the successful production of specific monoclonal antibodies.

Protocol 6. ELISA assay

Equipment and reagents

- Spectrophotometer with 450 nm filter
- Multiwell pipette (e.g. Gilson)
- Anti-mouse, anti-rat, or anti-rabbit antibody conjugated to horse radish peroxidase (HRPO) (Sera-Lab)
- Phosphate-buffered saline (PBS), pH 8.0
- PBS, containing 0.05% Tween 20.
- PBS, pH 8.0, containing 3% bovine serum albumin (BSA) or 3% skimmed milk powder.
- Immulon 2 flat-bottomed microtitre plates (Dynex Technologies)
- Tetramethylbenzidine hydrogen peroxide (TMB)
- 0.5 M H_2SO_4

A. *Preparation of antigen plates*

1. Coat immulon 2 ELISA plates with soluble antigen (proteins or peptides) as aliquots of 50 μl per well in phosphate buffer, pH 8.0, each containing 10–100 ng of antigen.
2. Incubate the plates at 4°C overnight.
3. To each well, add 200 μl of PBS, pH 8.0, containing 3% bovine serum albumin or 3% skimmed milk to prevent non-specific binding of the antibodies present in the culture supernatant.

B. *Preparation of antigen capture plates*

1. Coat the plates with a suitable capture antibody or lectin as aliquots of 50 μl diluted in PBS to a concentration of 5–10 μg ml^{-1}.
2. Incubate the plates at 4°C overnight.
3. Wash each well three times with PBS-0.05% Tween 20.
4. To each well, add 200 μl of PBS, pH 8.0, containing 3% BSA or 3% skimmed milk to prevent non-specific binding to the capture antibody.
5. Wash each well three times with PBS-0.05% Tween 20.
6. Add antigen in PBS to the plates at a concentration 1 μg ml^{-1}.
7. Incubate the plates at 4°C overnight, or at room temperature for 2 h.

C. *Assay protocol*

1. Wash each well of the coated plates three times with 200 μl of PBS containing 0.05% Tween.
2. Add 50 μl samples of test supernatant to each well in duplicate.
3. Incubate for 1 h at room temperature.
4. Wash each well three times with PBS-0.05% Tween 20.
5. Add 50 μl of HRPO-labelled second antibody (diluted in PBS-Tween according to a pre-determined optimum concentration) to each well.
6. Wash each well three times with PBS-0.05% Tween 20.

Protocol 6. *Continued*

7. Add 50 μl TMB to each well. Incubated the plates for 20 min at ambient temperature. A blue colour develops in the positive wells.
8. Add 50 μl 0.5 M H_2SO_4 to each well. The blue colour is converted to yellow.
9. Read the absorbance at 450 nm in a spectrophotometer.

Protocol 7. Radioimmunoassay

Equipment and reagents

- Gamma counter
- An anti-mouse, anti-rat, or anti-rabbit second antibody as appropriate, iodinated with ^{125}I (Nycomed Amersham)
- Phosphate-buffered saline (PBS), pH 8.0
- PBS, pH 8.0, containing 3% bovine serum albumin (BSA) or 3% skimmed milk
- PBS, pH 8.0, containing 3% BSA
- PVC multiwell flexible plates (Dynex Technologies)

A. *Preparation of antigen plates*

1. Coat PVC multiwell plates with soluble antigen (proteins or peptides) as aliquots of 50 μl per well in PBS, pH 8.0, each containing 10–100 ng of antigen.
2. Incubate the plates at 4°C overnight.
3. To each well, add 200 μl of PBS, pH 8.0, containing 3% bovine serum albumin or 3% skimmed milk to prevent non-specific binding of the antibodies present in the culture supernatant.

B. *Assay protocol*

1. Wash each well three times with 200 μl of PBS containing 3% BSA.
2. Add 50 μl samples of antibody-containing culture supernatants (or purified antibody) to each of the multiwell plates, usually in duplicate.
3. Incubate for 1 h at room temperature.
4. Wash each well three times with 200 μl of PBS containing 3% BSA.
5. Add 50 μl of ^{125}I-labelled secondary antibody (10^5 c.p.m. per 50 μl in PBS containing 3% BSA) to each well.
6. Incubate for 1 h at room temperature.
7. Discard the radioactive supernatant by inverting the plate over a sink designated for aqueous radioactive waste disposal.
8. Wash each well three times with 200 μl of PBS containing 3% BSA.
9. Measure the radioactivity in each well, representing antibody bound to the antigen, using a gamma counter.

4.3 Hybridoma production

Once the immunization regime is completed and the test bleeds show specificity of binding to the desired antigen, the fusion can be performed. Although methods vary depending on the animal and route of immunization, essentially the basis of hybridoma production is identical. Primed B cells are fused with an immortal cell line from the same species, and grown in a selection medium which results in the selection of hybridomas secreting the required antibody (*Figure 2*).

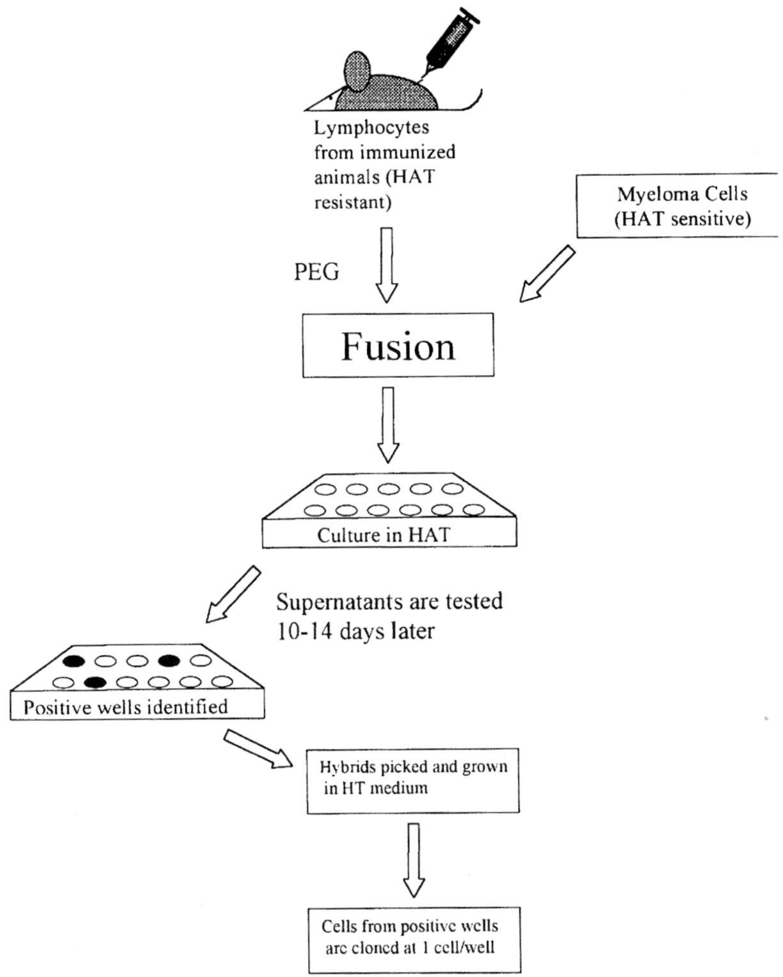

Figure 2. Fusion protocol for mice and rats.

Protocol 8. Cell fusion and cloning (mouse or rat cells)

Equipment and reagents

- Benchtop centrifuge
- Tissue-culture microscope
- Haemocytometer
- Spleen cells from immunized rat or mouse, or mesenteric lymph node cells (when immunized via the Peyer's patches)
- Mouse myeloma line or rat myeloma cell line (e.g. rat myeloma Y3-Ag1.2.3, usually grown in spinner flasks)
- Dulbecco's Modified Eagle's Medium (DMEM) containing glucose (4 g l^{-1}), bicarbonate (3.7 g l^{-1}), glutamine (4 $\times$ 10^{-3} M), penicillin (50 U ml^{-1}), and neomycin (100 μg ml^{-1})
- Fetal calf serum (FCS), inactivated by heating for 45 min at 56°C
- Feeder cells for fusion cultures: three hours to one day before fusion, dilute 2–4 $\times$ 10^6 rat fibroblasts in 10 ml of DMEM, previously irradiated with 30 Gy (3000 rad) of X- or γ-rays, with 90 ml of DMEM containing 10% FCS, and plate in 1 ml aliquots into four 24-well plates
- HAT selection medium:
  Prepare 100 $\times$ HT by dissolving 136 mg of hypoxanthine and 38.75 mg of thymidine in 100 ml of 0.02 M NaOH. Sterilize by filtration through a 0.45 μm membrane filter, and store at −20°C.
  Prepare 100 $\times$ aminopterin (A) by dissolving 1.9 mg of aminopterin in 100 ml of 0.01 M NaOH. Sterilize by filtration through a 0.45 μm membrane filter, and store at −20°C. (N.B. Aminopterin is toxic and should be handled with care.)
  HAT medium is prepared by adding 2 ml of HT and 2 ml of A to 200 ml of DMEM containing 20% FCS.
- HT medium: add 1 ml of HT to 100 ml of DMEM containing 10% FCS
- PEG solution: weigh 50 g of polyethylene glycol (MW 1500) into a capped 200 ml bottle, add 1 ml of water, and autoclave for 30 min at 120°C. Cool to about 70°C, add 50 ml of warmed DMEM, mix, and after cooling to room temperature, adjust the pH to 7.2 with 1M NaOH. Store as 1 ml aliquots at −20°C

A. *Fusion protocol*

1. Centrifuge exponentially growing rat or mouse myeloma cells in 50 ml aliquots for 3 min at 400 *g*. Wash twice by resuspension in serum-free DMEM. Count in a haemocytometer, and resuspend in serum-free DMEM at a concentration of 1–2 $\times$ 10^7 cells per ml.
2. Kill immune animals by exsanguination under anaesthetic (e.g. 3% halothane in oxygen), and open the abdominal cavity. Remove spleen or mesenteric nodes by blunt dissection.
3. Disaggregate spleen or nodes by forcing through a fine stainless steel mesh into 10 ml of serum-free DMEM, using a sterile syringe plunger.
4. Centrifuge cells for 3 min at 400 *g*. Wash twice in serum-free DMEM, and resuspend in 10 ml of the same medium.
5. Count viable lymphoid cells in a haemocytometer. Spleens from immune mice yield around 1 $\times$ 10^8 cells, spleens from rats 3–5 $\times$ 10^8 cells, and the mesenteric nodes of rats 1–2 $\times$ 10^8 cells.
6. Mix 1 $\times$ 10^8 viable lymphocytes with 5 $\times$ 10^7 rat or mouse myeloma cells in a 10 ml sterile capped tube, and centrifuge for 3 min at 400 *g*.
7. Pour off the supernatant, drain carefully with a Pasteur pipette, and release the cells by gently tapping the tube on the bench.
8. Stir 1 ml of PEG solution, pre-warmed to 37°C, into the pellet over a

period of 1 min. Continue mixing for a further minute by gently rocking the tube.

9. Dilute the fusion mixture with DMEM (1 ml over a period of 1 min, a further 1 ml over a period of 30 s, and finally 5 ml over 1 min).
10. Centrifuge for 3 min at 400 *g*, then resuspend the cells in 100 ml of HAT selection medium. Plate 1 ml aliquots into four 24-well plates seeded with irradiated fibroblasts.
11. Incubate the plates at 37°C in 5% CO_2. Check daily for presence of hybrids (usually visible after 6–14 days).
12. Once the positive fusion wells have been identified, pick out the individual colonies with a Pasteur pipette into 1 ml of HT medium in a sterile 24-well tissue culture plate, and split when cells become confluent.

B. *Cloning protocol*

1. Seed irradiated rat fibroblasts in DMEM–10% FCS, at 5×10^3 cells per well, into 96-well plates for use the next day.
2. Centrifuge cells from at least two wells of a 24-well plate that contain confluent layers of hybridoma cells. Count the number of cells, and dilute to give 50 cells in 20 ml of DMEM plus 10% FCS.
3. Plate 0.2 ml aliquots of hybridoma cells into each of the 96 wells.
4. Examine the plates 5–10 days later, and screen those wells containing visible single colonies.
5. Pick positive clones into 24-well plates, expand, and freeze under liquid nitrogen.
6. Re-clone the colonies producing the greatest amount of specific antibody.

Protocol 9. Cell fusion and cloning (rabbit cells)

Equipment and reagents

- Benchtop centrifuge
- Tissue-culture microscope
- Haemocytometer
- Rabbit plasmacytoma cell line 240E-1 in log phase.
- Spleen cells or popliteal-node cells from an immunized rabbit (*Protocols 4, 5*).
- 50% polyethylene glycol (PEG) 4000.
- Hypoxanthine, Aminopterin, Thymidine selection medium (*Protocol 8*).
- RPMI 1640 medium enriched with following additions: amino acids, non-essential amino acids, pyruvate, glutamine, vitamins, Hepes, gentamicin, penicillin, streptomycin, fungizone (all components from GIBCO and used at concentrations as suggested by the supplier), 50 mM 2-mercaptoethanol, 7.5% sodium bicarbonate, and 15% FCS; adjusted to pH 7.2with 1 N HCl or 1 N NaOH.

Protocol 9. *Continued*

A. *Fusion protocol*

1. Centrifuge rabbit plasmacytoma cell line 240E-1 in log phase in 50 ml aliquots for 3 min at 400 *g*, wash twice by resuspension in serum-free RPMI 1640, count in a haemocytometer, and resuspend in this medium to 1–2 $\times$ 10^7 cells per ml.
2. Kill immune rabbit by exsanguination under anaesthetic (e.g. 3% halothane in oxygen), and open the abdominal cavity. Remove spleen by blunt dissection.
3. Disaggregate spleen by injection with sterile PBS, 'ballooning' the spleen, or disaggregate nodes by forcing through a fine stainless steel mesh into 10 ml of serum-free DMEM, using a sterile syringe plunger.
4. Centrifuge cells for 3 min at 400 *g*, wash twice in serum-free RPMI 1640, and resuspend in 10 ml of the same medium.
5. Count viable lymphoid cells in a haemocytometer.
6. Mix 1.5–3 $\times$ 10^8 viable lymphocytes with the rabbit fusion partner at a ratio of 2:1 in a 10 ml sterile capped tube, and centrifuge for 3 min at 400 *g*.
7. Pour off the supernatant, drain carefully with a Pasteur pipette, and then release the cells by gently tapping the tube on the bench.
8. Stir 1 ml of PEG 4000 solution, pre-warmed to 37 °C, into the pellet over a period of 1 min. Continue mixing for a further 1 min by gently rocking the tube.
9. Dilute the fusion mixture with RPMI 1640 (1 ml over a period of 1 min, a further 1 ml over a period of 30 s, and finally 5 ml over 1 min).
10. Centrifuge for 3 min at 400 *g*, then plate the cells in 48-well plates at 2 $\times$ 10^5 lymphocytes in 0.5% RPMI 1640 medium containing 15% FCS.
11. After 72 h, add 0.5 ml of 2$\times$ concentrated HAT medium. Change 50% of the medium every 4–5 days.
12. Clones can usually be seen after 2–5 weeks.
13. Screen culture supernatant for specific antibodies using ELISA (*Protocol 6*) or radioimmunoassay (*Protocol 7*).

B. *Cloning protocol*

Follow instructions in *Protocol 8*.

4.4 Analysis of fusions

Whatever the choice of screening method, it is desirable that the assay be quick, reliable, and specific. The supernatant from the fusion wells can be screened using many different procedures: recognition of virus infected cells and monitoring by fluorescence-activated cell sorting (FACS), recognition of virus antigen by *in situ* staining, EIA, RIA, or any other appropriate method. In the first instance it is essential to identify rapidly clones producing antibody to the immunising antigen, in order to identify those wells which contain hybridomas producing the required antibody. Hybrids from these positive wells can then be isolated, grown, re-tested, and finally cloned. The supernatant from these wells can be harvested to provide antibody.

5. Isotyping

Antibodies can be separated into different classes (IgM, IgA, IgE, and IgG) and subclasses (IgG_1, IgG_{2a}, IgG_{2b}, IgG_{2c}) for rats, on the basis of the structure of their heavy chains. Isotyping kits are commercially available from several sources (e.g. BioRad Laboratories, Boehringer Mannheim). This separation can be achieved using antisera specific for antigens on the heavy chains, which are specific for a particular isotype or subclass. The different classes or subclasses will determine the affinity for important secondary agents such as protein A and protein G. Knowledge of the subclass of an antibody is required in order to decide on the best method of purification.

6. Purification

Purification of a monoclonal antibody from cell culture usually requires a combination of techniques. Each technique should utilize a particular physical or chemical attribute of the mAb that allows it to be separated from molecules lacking the phenotype in question. The molecular weight, molecular charge, and hydrophobicity can be used to advantage in sample preparation steps such as precipitation, and in subsequent purification methods (*Table 2*).

Table 2. Antibody purification methods

Antibody	Protein A	Protein G	DEAE(DE52)	Affinity	Gel filtration
Rat IgG_1	−	+	+ (0.055 M)	+	
Rat IgG_{2a}	−	++++	+ (0.0175 M)	+	
Rat IgG_{2b}	−	++	+ (0.0175 M)	+	
Rat IgG_{2c}	+	++	+ (0.0175 M)	+	
Rat IgA	−	+	−	+	+
Rat IgM	−	−	−	+	+
Mouse IgG_1	+	++++		+	
Mouse IgG_{2a}	++++	++++		+	
Mouse IgG_{2b}	+++	+++		+	
Mouse IgG_3	++	+++		+	
Rabbit (polyclonal)	++	++		+	

Monoclonal antibodies are unique in the possession of an Fc fragment which binds specifically and reversibly to protein G and protein A molecules. These can be linked to a matrix, so that affinity chromatography techniques can be used (*Protocol 3*).

References

1. Kohler, G., and Milstein, C. (1975). *Nature (London)* **256,** 495.
2. Dean, C. J., Styles, J. M., Gyure, L. A., Peppard, J., Hobbs, S. M., Jackson, E., and Hall, J. G. (1984). *Clin. Exp. Immunol.*, **57,** 358.
3. Galfre, G., and Milstein, C. (1981). In *Methods in enzymology* (ed. Langone, J. J., and Van Vunakis, M.). Vol. 73, p. 3. Academic Press, New York.
4. Bazin, H. (1982). In *Protides of the biological fluids, 29th colloquium* (ed. Peters, H.). p. 615. Pergamon, New York.
5. Edwards, P. A. W., Smith, C. M., Neville, A. M., and O'Hare, M. J. (1982). *Eur. J. Immunol.,* **12,** 641.
6. Anderson, D. V., Tucker, E. M.., Powell, J. R., and Porter, P. (1987). *Vet. Immunol. Immunopathol.,* **15,** 223.
7. Flynn, J. N., Harkiss, G. D., and Hopkins, J. (1989). *J. Immunol. Meth.*, **121,** 237.
8. Richards, C. M., Aucken, H. A., Tucker, E. M., Hannant, D., Mumford, J. A., and Powell, J. R. (1992). *Vet. Immunol. Immunopathol.,* **33,** 129.
9. Spieker-Polet, H., Sethupathi, P., Yam, P.-C., and Knight, K.L. (1995). *Proc. Natl. Acad Sci. USA,* **92,** 9348.

8

Antiviral testing

J. S. OXFORD, L. S. KELLY, S. DAVIES and R. LAMBKIN

1. Introduction

The clinical effectiveness of acyclovir, and more recently penciclovir, against herpes simplex type I and II infections (1) has led both to great expectations of further discoveries of antiviral compounds and to a change in the scientific direction of clinical diagnostic laboratories. Most such laboratories are at present used primarily to diagnose viral infections, either by culture or serology, or, more often, by detection of viruses using rapid molecular techniques such as PCR. The advent of the new generation antivirals such as protease inhibitors of HIV-1 (2), and neuraminidase inhibitors of influenza (3–5), together with new molecular methodologies, now requires clinical laboratories to quantify virus, usually by PCR or by molecular probing. Also, the therapeutic benefit of antiviral therapy has been enhanced by the use of PCR to analyse viral genes and search for drug-resistant mutations (6). Nowhere is this more important than with HIV, where, on the one hand, a choice of antivirals exists, and on the other hand the introduction of combination chemotherapy has led to a substantial increase in demand for scientific information on which to base clinical decisions.

In addition, research into new antivirals is now more than ever before focused in two regards. Firstly, X-ray crystallographers can refine chemical structures that are known to bind to key viral target enzymes, for example the HIV-1 protease (7, 8). Secondly, combinatorial chemistry can greatly increase the number of chemicals to be tested. It is now not unusual to screen hundreds of thousands of molecules against a key viral target. So the new technology is large-scale screening against viral enzymes, using PCR and ELISA techniques. But the discovery of a lead compound that inhibits the function of a viral enzyme is still a long way from an inhibitor that can penetrate the plasma membrane of a cell and, once there, interrupt viral replication. It may be even further away from a drug which, taken orally, can be absorbed, have high bioavailability, and still inhibit viral replication in a selective manner.

It has always been considered in the past that an *in vitro* screen of 100 000 molecules could lead to approximately 100 with *in vitro* activity, 20 with *in vivo* activity, and perhaps one that could enter clinical trials. The role of the virology laboratory will be even more crucial in the future, to test molecules in tissue culture, and finally in defined laboratory animals. Alongside the developments in scientific discovery and analysis of antiviral compounds, the important aspect of clinical testing has also changed, and has become more analytical and more precisely controlled. All clinical trials are now strictly governed by international regulations. This includes independent monitoring of the trial, whether phase I (first into humans, usually involving a low number of volunteers, e.g. ten), phase II, perhaps involving 200–400 healthy volunteers, phase III, involving patients with the viral disease in question, or phase IV, which allows more careful monitoring in special, perhaps more 'at risk' groups.

Most antivirals are discovered in specialized commercial chemistry laboratories, not always in the largest groups of the pharmaceutical industry. More often there is extensive formal collaboration of pharmaceutical companies with academic groups who have a very specialized interest, for example, in HIV integrase as a target enzyme. Finally, contract research organizations undertake small-to-medium scale cell culture screening for antivirals, and can maintain small focused research teams. Therefore, in the last decade, the science of antivirals has come of age, and the cultivation of viruses *in vitro* is a central theme of any virology research programme. Antiviral screening has immense practical consequences, and will undoubtedly be a key element in research and clinical laboratories and in clinical testing units over the next decade. *In vitro* and *in vivo* screens will act as a gate, at which point decisions will be made on the progress of a drug to the final hurdle of the clinical trial.

1.1 Inhibitors of critical viral enzymes

Many viruses encode a vital enzyme that differs from a cellular counterpart, for example, herpes DNA polymerase and thymidine kinase, or influenza neuraminidase. In addition, many viruses have absolutely unique enzymes with no known cellular equivalents, such as the RNA replicases and transcriptases of RNA genome viruses. Furthermore, many viruses control their transcription using strong promoter sequences, and encode unique regulatory proteins. These viral enzymes and proteins are the most recent targets of antiviral chemotherapists. In fact, several successful antiviral compounds act on viral enzymes, such as acyclovir against herpes DNA polymerase, and AZT and other dideoxynucleoside analogues against HIV reverse transcriptase. But these compounds were discovered by screening in culture systems and not by using enzyme inhibition assays.

As an example, herpes simplex virus type 1 encodes its own thymidine kinase enzyme. Thymidine kinase (TK) is an important enzyme in the pyrimidine salvage pathway, rescuing thymidine from catabolic processes. The human cytosolic isoenzyme catalyses the monophosphorylation of the 5′-OH group of thymidine, using ATP as the phosphate donor. Thymidine monophosphate is then converted to thymidine triphosphate and used for DNA synthesis. The TK enzyme encoded by HSV-1 has much wider substrate specificity than its human equivalent, and it will metabolize derivatives of the nucleosides, for example, guanosine derivatives such as acyclovir (ACV) and ganciclovir. The difference in substrate specificity between the host cell-encoded enzyme and the viral-encoded enzyme accounts for the virustatic effect of compounds such as acyclovir (9). ACV is a monophosphorylated prodrug that requires phosphorylation to the triphosphate form *in vivo*. In this form it is then incorporated into DNA and causes chain termination. The phosphorylation step occurs using the viral-encoded TK, and therefore chain termination is restricted to herpes-infected cells. An understanding of the molecular basis underlying this different substrate specificity could be the key to future rational compound design, and could reduce the need for very large scale 'shotgun' style screening of potential antiviral compounds.

Recombinant DNA technology can produce quantities of most viral proteins and, together with ELISA technology, antiviral chemotherapists now have very powerful screening methods. Although this may be applicable to certain transcription factors and particular viral enzymes such as influenza neuraminidase or HIV protease, other potentially important viral enzyme targets are more complex. Thus the influenza RNA replicase complex has at least two proteins, and there is still a lack of X-ray crystallography data to enable inhibitors to be designed. In this case the virus must be cultivated in large quantities (1 g batches for example), and then disrupted with mild detergent to activate RNA transcriptase and replicase activity. Undoubtedly these assays will be developed further in the next few years in the quest for important new antiviral compounds, and also adapted to screen the millions of drugs in compound libraries.

2. The role of cell culture in testing antiviral compounds

Cell culture techniques (Chapter 1), with the notable exception of novel methods to grow human T cells in suspension, have changed little in a fundamental way in the years since Enders received the Nobel Prize in 1954 for cultivation of poliovirus from non-neural tissue. But there is now a realization that cells can be cross-contaminated or quietly infected with mycoplasma or latent viruses. To some extent, cell culture is still a 'green-fingered' scientific endeavour. Most easily cultivated are continuous cell lines that have either been transformed spontaneously or in the laboratory by oncogenic viruses. Typical examples of attached cells are MDCK (Madin Darby Canine Kidney), Vero, and HeLa, which are used for cultivation of influenza, polio, and herpes respectively. But these cells carry dangers of their own, because they can easily cross-contaminate other cell lines and become the dominant cell type.

In contrast, semi-continuous cells have a limited lifespan in culture of approximately 40 or 50 cycles; examples are the MRC-5 and WI-38 cells originating from human embryos. These cells are significantly more difficult to grow in the laboratory. Finally human C8166 cells or PBMCs are cultivated and maintained by PHA and IL2 stimulation in suspension. C8166 are human T cells that have been transformed by a human retrovirus, but no longer carry the full retrovirus genome.

There has been an important innovation in the introduction of 96- and 48-well tissue-culture plates, and also operationally in the use of class II safety cabinets, which give some protection to the scientist and also some protection to the cells against contaminating bacteria. The use of 96-well plates that are not only suitable for tissue culture because of their coating, but also have suitable optical characteristics to allow the objective determination of colorimetric changes arising from viral infection (e.g. the metabolization of the tetrazolium salt) by an ELISA plate reader, have substantially enhanced the rate of drug throughput for antiviral testing.

2.1 *In vitro* analysis

Most antiviral drugs exert an effect on virus attachment or release from an infected cell or, alternatively, penetrate the cell and inhibit a viral virus enzyme or a stage in replication. A so-called virustatic assay is devised to quantify their effect *in vitro*. There are four guiding principles:

- Carefully consider the cell line, because some antivirals only exert an effect in particular cells.
- Vary the challenge dose of virus, bearing in mind that no existing antivirals can withstand an overwhelmingly high input of virus.
- Allow 30 minutes or so for the drug to penetrate into the cell and reach equilibrium.
- Test varying concentrations of drug which have been pre-tested for absence of a direct toxic effect on the cells.

The primary purpose of the assay is not to reproduce a human infection, but to maximize the chance of discovering a 'lead' compound with virus inhibitory effects. Briefly, a particular cell line that is known to be sensitive to infection by the virus of interest is incubated for a short period of time with the test compound. After a predetermined period, the cells are challenged with the virus at a specific infective titre. If the compound possesses anti-viral effects, the ability of the virus to infect and/or replicate should be reduced. In all cell systems an appropriate control drug is used, that is, a compound known to be effective at reducing the replicative activity of the virus of interest, for example, acyclovir for herpes simplex 1, and zidovudine (AZT) for HIV-1. Virus replication is quantified after incubation for 2–7 days. The mechanism by which this quantification is achieved will vary, depending on

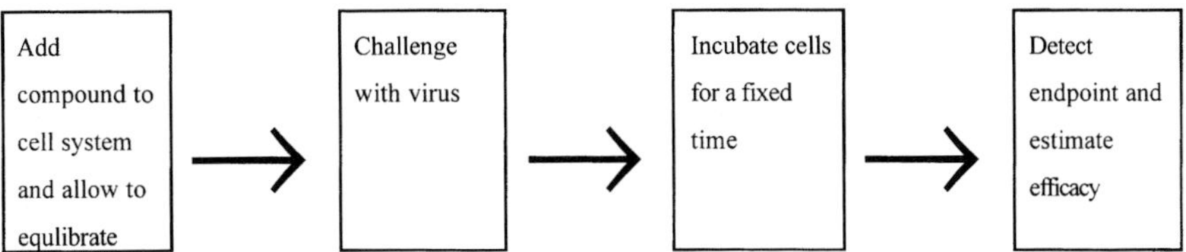

Figure 1. Schematic diagram of a virustatic assay.

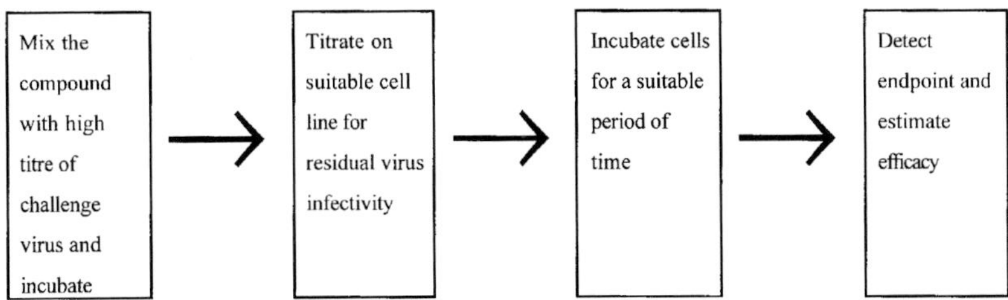

Figure 2. Schematic diagram of a virucidal assay.

the virus being tested. The haemagglutination assay is used to detect influenza and parainfluenza viruses, the MTT colorimetric assay is used for HSV-1 and RSV, and an enzyme-linked immunoassay (ELISA) is used to detect HIV p24 production (*Figure 1*).

It is possible that compounds could act like disinfectants and destroy virus on contact, perhaps at mucosal surfaces during sexual transmission, or in the upper respiratory tract or throat. The assay system of choice here is the virucidal one. The basic principal is to incubate varying concentrations of the test compound with a high infectious titre of virus for 30 minutes at 37°C. Any residual virus is quantified by titration in an indicator cell line (*Figure 2*). The compound is pre-screened for cellular toxicity and tested at non-toxic concentrations. The use of a high titre of challenge virus at the incubation step is to allow the subsequent dilution of the drug–virus mixture, whereby no significant concentration of the drug is carried over into the indicator cell line, where it might otherwise have a virustatic effect.

In all cell systems an appropriate control is used, that is, a compound known to be effective in reducing the activity of the virus of interest, for example, acyclovir for herpes simplex virus 1, and zidovudine for HIV-1 in virustatic assays (10, 11), or sodium deoxycholate for HIV-1 in virucidal assays. *Table 1* contains details of common human viruses, and suitable cell lines for antiviral testing.

3. The design of the anti-viral testing laboratory

A standard drug-screening laboratory will resemble most tissue-culture laboratories, with its most important function to maintain and allow the

Table 1. Human viruses and the cell lines commonly used for cell culture

Virus	Cell lines	Incubation time (days)	Effect on cells	ATCC code
Adenovirus	HeLa	4	CPE	VR-913
Coronavirus	HEK MRC-5 WI-38	7	CPE	VR-740
Herpes simplex types 1 and 2	BHK-21 Vero	2–5	CPE	VR-260
Varicella zoster	MRC-5	4–7	Gradual spreading of focal type CPE from cell to cell	VR1367
Human rhinovirus	MRC-5 WI-38 HeLa	1–7	CPE	VC-521
Influenza A and B	Chick embryo MDCK Vero MRC-5	3–7	Culture fluid agglutinates RBC	Various
Parainfluenza	MkK cells MKHe HeLa	5	Culture fluid agglutinates RBC	VR-93
Respiratory syncytial virus	HEp-2 MRC-5 Vero	5–7	CPE; massive syncytia in culture	VR-1401
Rotavirus	MA-104 CV-1	5–7 days at 37°C	CPE but variable.	VR-2018

handling of cell cultures without cross-contamination or contamination with micro-organisms. The laboratory will be divided into two main areas, to reflect the two types of procedures carried out within the space. 'Non-infected procedures' describes all of the essential operations necessary to maintain non-contaminated cell lines, and to produce sufficient cells for *in vitro* assays. 'Infected procedures' describes all the operations in which viruses and contaminated cell lines are handled. It is essential to keep these two activities separate within the laboratory. This can easily be achieved with duplication of equipment, and by physical separation to produce a non-infected gradient within the laboratory. Every effort must be made to keep equipment separated. Where possible, everything from laminar flow cabinets to pipettes should be designated for either 'non-infected' or 'infected' operations. Therefore an antiviral screening laboratory will have substantially more equipment than a standard tissue culture laboratory. Both the 'non-infected' and 'infected' areas will have:

1. **Laminar flow cabinets**. For a containment level II laboratory, two class II cabinets (one for 'non-infected' and one for 'infected' procedures) and

possibly (depending on space) one class I cabinet. For a containment level III laboratory, one class II cabinet (for 'non-infected' procedures) and one class I cabinet (for infectious material).

2. **CO_2 incubators**. There should be at least two CO_2 incubators, for infectious and non-infectious material. The most suitable incubators are designed to reduce the available surfaces to harbour potential microbial contaminants. In addition they should be water-jacketed to ensure even temperature control. The CO_2 supply should have an in-line microbial filter to prevent potential contamination.
3. **Storage space for equipment and solutions**. If possible, this should be in another room to help prevent clutter and reduce any risk of contamination.
4. **Instrument and equipment benches**. It is not always possible to duplicate all pieces of equipment, due to space and economics, and if this is the case then further measures can be taken, such as regular cleaning of centrifuges, water baths, and plate shakers, to ensure that cell lines remain uncontaminated. The ideal benching should be manufactured from solid phenolic resins, and not laminates. Damage to laminated benching is common, and these can provide ideal locations for harbouring microbial contamination. Bench cabinets and fittings should be fitted to the metal frame of the bench structure itself, and not be free-standing. Ideally this should enable at least 6–10 cm clearance between the floor and under-bench fittings, allowing thorough cleaning to prevent the accumulation of detritus.
5. **Refrigerators and freezers**. It is also advisable to have refrigerators and freezers separated, or at the very least, designated areas within +4°C, −20°C, and −70°C storage. Emergency backup should be fitted to the −70°C freezer, sufficient to maintain the temperature within range for at least 48 hours.
6. **Alarm systems**. To monitor all temperature-controlled equipment, computerized dialling of the emergency telephone numbers should be considered as a worthwhile investment. The downtime of the laboratory should a −70°C freezer fail, and the resulting cost in lost working time, would undoubtedly be much greater.
7. **Liquid nitrogen storage**. Again, every effort must be made to keep the liquid nitrogen storage orderly and separated. A logging system should be introduced, and efficient procedures must exist to ensure that the liquid nitrogen does not 'run dry'. It is worthwhile considering contingency storage of essential cell lines in the −70°C freezer. Although long-term viability of cells not stored below the 'glass point' is poor, medium term storage (1–2 years) is satisfactory.

Other equipment that would be required includes:

- an inverted microscope
- a normal microscope

- a sensitive balance suitable for measuring as little as 10 mg
- magnetic stirrers
- an ELISA plate reader capable of being linked via a serial cable to a personal computer
- personal computer
- computer printer
- an ELISA plate washer
- a low-speed desktop centrifuge with suitable carriers varying in size from Eppendorf tubes (0.5 ml to 1.5 ml), Universal tubes (20 ml), Falcon tubes (50 ml), and ideally 48-well and 96-well plate-carriers
- small benchtop autoclave capable of automated functioning

There are numerous factors influencing the positioning of equipment. Laminar flow cabinets must be within easy access of refrigerators, freezers, and incubators to allow efficient working within the laboratory. The safety of the individual within the laboratory is of paramount importance. Positioning of the laminar flow cabinets must be carefully considered, to maintain the airflow in the cabinet and also to protect the person. Class I cabinets offer good protection to the individual, with airflow extracted away from the person working; however, there is an increased risk of contamination of the cell culture. Class II cabinets offer greater protection to the cells or reagents within the cabinet than does a class I cabinet, but offer less protection to the individual. The cabinet must be positioned away from doors and windows, and not in busy laboratories or near vents or air-conditioning units, as all of these will have an effect on the overall airflow of the laboratory, and therefore a potentially detrimental effect on the airflow of the cabinet itself.

Safety considerations are especially important when working within a containment level III laboratory, where extra thought must go into the positioning of laboratory equipment and the organization of space. The appropriate national guidelines for category III work must be followed; only competent and experienced personnel must work within the laboratory.

Most antiviral screening is carried out in class II laboratories using viruses such as influenza A and B, herpes simplex, cytomegalovirus, rhinovirus, and respiratory syncytial virus. Most of these viruses are well adapted to cell culture, and are not highly pathogenic to laboratory personnel. However, there is an increased interest in the evaluation of drugs against unadapted field- or street-viruses, and safety aspects of the laboratory may have to be re-evaluated. Laboratory infections have been documented with unpassaged influenza viruses, both from infected embryonated hens eggs, and from infected animals. Therefore extra attention should be given to staff experience, and a possible vaccination strategy reassessed on a year-by-year basis. There would appear to be less likelihood of laboratory infection with herpesviruses, and no particular worry with rhinoviruses, because of their common community occurrence throughout the year, and the mild clinical effects exerted.

Drug screening against HIV must be carried out in a category III laboratory. The key element here is staff experience and training. Only a single laboratory-acquired infection has been documented, and nowadays some laboratories even cultivate HIV in a class II cabinet, albeit a modern and frequently sterilized one. Most category III laboratories also maintain supplies of, or rapid access to, nucleoside analogues and protease inhibitors, for instant use under medical supervision for prophylactic use in the event of occupational exposure to HIV.

4. Testing procedures

The effectiveness of an antiviral compound against a virus can be determined by quantifying the effect of the compound on the virus replication in tissue culture. The host cell system, the infectious titre of the challenge virus and therefore its multiplicity of infection (MOI), and the period of incubation before the endpoint of the experiment is determined, will all affect the sensitivity of the system. The two traditional virustatic methods for determining the sensitivity of a virus to an anti-viral compound are the plaque reduction assay and the virus-yield reduction assay. Both these methods can also be used to test putative drug-resistant mutants as well. In this case, a virus would be tested against a range of compound concentrations known to be effective in parallel with a virus of known (and normal) sensitivity. Nucleotide sequencing of the mutant may then be conducted to identify the mutation.

In a viral plaque assay, a confluent monolayer of a suitable cell line, either pre-incubated or not with the drug, is infected with the virus of interest. The virus inoculum is removed after a period of time sufficient for the virus to enter the target cells, and a solid overlay of nutrient medium is added. As no culture supernatant is present, virus infection of other cells can only occur by release of virus and infection of neighbouring cells. In this way small regions of necrosis or cell destruction occur in the cell monolayer, and these are known as plaques (*Figure 3*). Depending on the virus, these plaques can vary greatly in size, and are revealed after incubation by removal of the overlay and staining with crystal violet or other suitable cell-staining solution. The reduction in the number of plaques when incubated in the presence of the test compound can be used to measure the sensitivity of the virus to the compound.

A more labour-intensive and demanding assay system is the virus-yield reduction method. In this method, the virus is cultured in flasks in the presence or absence of the test compound. Suitable controls include virus alone with no compound in one flask, and a compound known to be effective against the virus in another. After a suitable period, usually between 2 and 7 days, samples of the tissue-culture supernatant are removed and titrated for

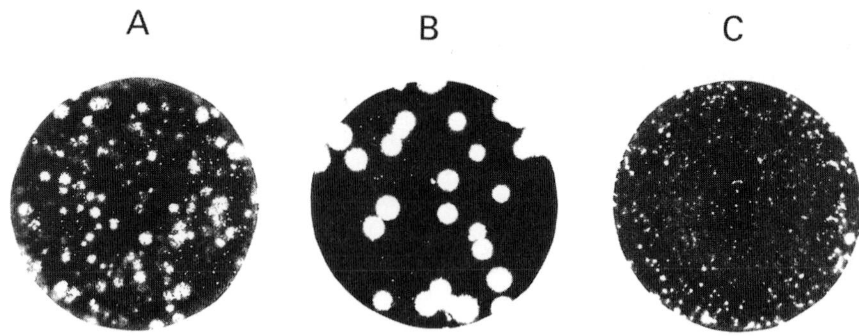

Figure 3. Variation in plaque morphology of influenza A virus isolates.

virus by plaque assay or other suitable technique, for example, haemagglutination assay for influenza.

These techniques have the disadvantages that they use a large amount of resources, both in staff and materials. As a result, they have a very low throughput, and are not suitable for the large-scale screening of antiviral compounds. Nevertheless, these two assays remain the cornerstone of the development of a new antiviral drug. The protocol for generic plaque-reduction assay is given below, and a protocol for a viral load reduction assay is given in the section on influenza antiviral testing.

Protocol 1. Plaque-reduction assay for testing of antiviral compounds

Equipment and reagents

- A suitable host cell line for the virus
- Minimum Essential Medium (Sigma, Life Technologies, ICN Flow, etc.)
- Class II safety cabinet
- Sterile plastic pipettes (Phillip Harris)
- Multiwell tissue culture plates (e.g. Nunc)
- Crystal violet 0.1% (w/v)
- Phosphate-buffered saline (PBS)
- Solution A: MEM 32 ml; bovine serum albumin (3.5%) 6.4 ml; DEAE-dextran (1%) 3.2 ml; antibiotics (penicillin, streptomycin: Life Technologies) 0.96 ml; Na_2CO_3 (4.4%) 12.8 ml; distilled water, 104 ml; trypsin 1 ml.
- Solution B: 160 ml 1% agar: autoclaved and then kept at 45 °C in a waterbath.

Method

1. Monolayers of cells are produced in suitably sized wells. The size of the well used is determined by the virus and the nature of the plaques produced.
2. Determine the challenge dose of virus by prior titration of a stock preparation on the cell line to be used.
3. Make serial dilutions of the antiviral compound.

4. Remove the supernatant from each well, and wash the monolayer twice with sterile PBS.
5. Add the serial dilutions of the antiviral compound to the plate, and allow to equilibrate for 1 h.
6. Remove the supernatant, and wash the monolayer twice with sterile PBS.
7. Add sufficient challenge virus to produce approximately 50 plaques (i.e. 50 pfu) after incubation.
8. Allow the virus is to adsorb for 1 h.
9. Mix solutions A and B together, and add the dilutions of the test compound to aliquots of the overlay to give the same final concentration as used earlier.
10. Distribute the overlay into each well.
11. Leave the dishes until the medium is solid, and then place them in the incubator at 37 °C.
12. Incubate for 2–7 days, depending on the virus
13. Remove the agar from each well, and stain the monolayer by the addition of 0.1% crystal violet.
14. Incubate the plates for 30 min at room temperature, and then gently wash with tap water.
15. Count plaques by visual examination.

4.1 *In vitro* testing of compounds for antiviral effect against influenza A

The embryonated hen's egg has been the traditional method for cultivation of influenza viruses. In this system, fertilized eggs half-way through their incubation period are infected with influenza virus by inoculation into the allantoic cavity. After incubation at between 35 °C and 37 °C for 18 to 48 hours (depending on the virus in question), the embryos are killed by chilling overnight at 4 °C, and the allantoic fluid containing the virus is harvested. However, in many countries, the use of embryonated hens' eggs is governed by the same regulations as those for laboratory animal welfare. In the United Kingdom, for example, it is necessary to possess a personal license issued by the Home Office, and to have undergone a training programme before hand. These restrictions, and the need to simplify and partly automate the process of influenza virus cultivation, has led to the development of tissue-culture techniques using multiwell plates. The MDCK cell line is an established method for the detection of the cytopathic effect of influenza in a tissue culture-based assay (12). Other cell lines are used for both clinical isolation of the virus and virus cultivation, but are much less sensitive than MDCK cells. Examples of additional cells are MRC-5 (13) and Vero cells (14).

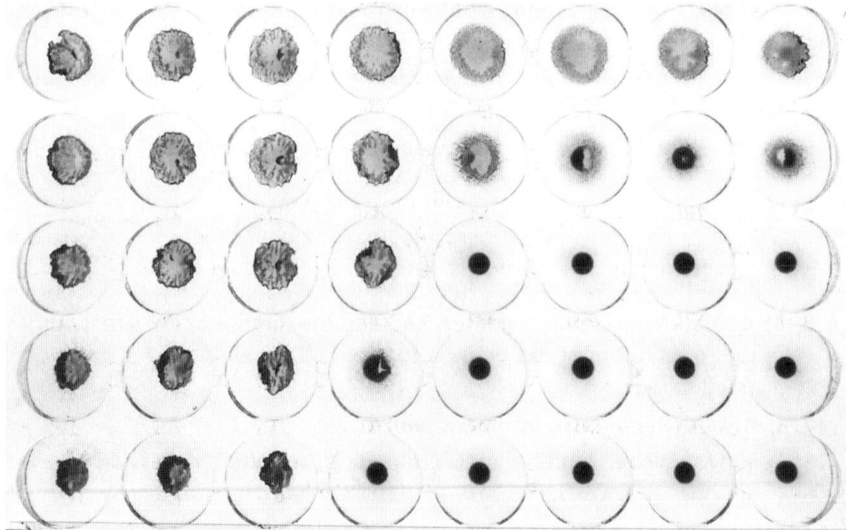

Figure 4. The haemagglutination assay.

There are several methods for detecting the endpoint in viral titration or antiviral assays of influenza. The simplest is to measure the haemagglutination titre (*Protocol 2*). This is a simple assay that relies on the ability of influenza virus to cross-link the sialic acid residues on the surface of red blood cells (turkey, chicken, and guinea pig cells are all used, depending on the type of influenza virus being titrated) that hold the cells in a diffuse matrix and prevent them from settling out of suspension. A 96-well V-shaped plate is used. At a particular dilution, there will be insufficient virus to cross-link the red blood cells, and they will settle. As the well is V-shaped, the cells will collect at the bottom. Viewed from above, this will appear as a small button in the centre of the well (*Figure 4*). However, observational skill is required, because at high concentrations of virus the matrix of cross-linked RBCs can collapse in on themselves, forming what appears to be a button. Gently tipping the plate, however, distinguishes true buttons from collapsed haemagglutination, as those cells not cross-linked will trickle forward to form a tear shape, whereas the collapsed haemagglutination will remain unchanged. In many laboratories, the haemagglutination assay remains the preferred method for detecting influenza virus growth. It is a quick and simple assay with high reliability. A companion assay (the haemagglutination inhibition assay) is still used as the international standard for measuring the efficacy of influenza vaccines, as required for licensing.

Protocol 3 describes a viral yield reduction assay for testing a putative influenza antiviral compound against influenza A virus, using amantadine as a control. This method would only be suitable for a small number of test compounds, as the labour input involved is high. If the number of compounds to be tested is greater, then multiwell plates can be used. The sensitivity of the assay tends to be reduced in multiwell plates, but the benefits for automating aspects of the work greatly outweigh this disadvantage. A 96-well plate assay (*Protocol 4*) is, in essence, similar to the assay in flasks, except that a lower volume of virus is used. Detecting the endpoint in this assay can be achieved by several methods. Measuring the quantity of influenza virus present in the supernatant in each well is practicable as long as the number of samples being tested is not too high. Alternatively, the level of virus-induced CPE can be measured. This technique is commonly used to determine the endpoint in antiviral assays of cytopathic viruses. If, as is the case with MDCK cells, the cell line is adherent, simply staining the remaining monolayer after incubation (usually 5–7 days) with crystal violet will allow visual determination of the endpoint. This can be partly automated by the addition of 50% methanol solution to the stained monolayer, thus solubilizing the crystal violet, and then measuring the resulting coloured solution in an ELISA plate reader.

Protocol 2. Titration of influenza virus using the haemagglutination assay

Equipment and reagents

- 96-well U- or V-bottomed microtitre plate
- 0.75% v/v chicken red blood cells (CRBCs) (Advanced Protein Products Ltd)
- Multichannel pipette (Anachem), or digital multidispensing pipetter (e.g. Jensen Seal-pipette)

Method

1. Dilute the virus, using a multi-channel pipetter and sterile disposable tips, in PBS across a 96-well U- or V-bottomed microtitre plate, final volume 100 μl.
2. Using a multidispensing pipetter, add 25 μl of 0.75% red blood cells (RBCs) to each well.
3. Mix the wells by repeated pipetting actions, using a multichannel pipetter and disposable yellow tips. Remove the first millimetre of the tip with scissors so that the aperture is large enough for shearing forces not to damage the RBCs.
4. Incubate the plates at room temperature for 1 h.
5. In wells that contain virus, the RBCs haemagglutinate, i.e. form a diffuse matrix, and do not settle in the bottom of the well, as the RBCs in the virus-free wells do.
6. The highest dilution of virus to cause the RBCs to haemagglutinate is 1 haemagglutinating unit (HAU). The titre of the virus preparation is expressed as HAU per ml.

Protocol 3. Virus yield reduction assay of a test compound against influenza A

Equipment and reagents

- Minimum Essential Medium containing 50 IU ml^{-1} penicillin, 50 $\mu g\ ml^{-1}$ streptomycin, and 2 mM L-glutamine (Sigma, Life Technologies, ICN Flow, etc.)
- Gilson pipette (100 μl) and sterile tips.
- 5 ml and 10 ml sterile plastic pipettes (Phillip Harris)
- Compound to be tested, at predetermined concentrations
- 25 cm^2 cell culture flasks (e.g. Corning, Falcon, Costar, etc.)
- Phosphate-buffered saline
- 0.05% TPCK trypsin (Worthington)
- Previously titrated influenza virus preparation (*Protocol 2*)
- Amantadine (test control substance) (Noveargis Pharmaceuticals UK Ltd)

Method

1. Produce confluent 25 cm^2 flasks of MDCK cells.
2. Remove the culture medium, and wash the cells twice with PBS.
3. Dilute the test compound to the required concentration in DMEM containing 2.5 $\mu g\ ml^{-1}$ of TPCK-treated trypsin, *but no fetal calf serum.**
4. Incubate the flask at 37 °C for 60 min to allow the cells and compound to equilibrate.
5. Dilute the influenza virus to approximately 128 HA units (10^5 $TCID_{50}$), and add 500 μl to each of the flasks.
6. Keep three control flasks: one with only culture medium and trypsin, one with only cell culture medium, and one with no compound but still challenged with the virus.
7. Challenge a fourth flask, containing a known efficacious concentration of amantadine hydrochloride, with the virus (positive control).
8. Incubate the flasks at 37 °C in a 5% CO_2 incubator.
9. Remove aliquots of tissue culture medium (50 μl) from the flask daily, and titrate by haemagglutination assay.
10. Compare the HA titres from the compound-treated and untreated flasks. A difference of greater than fourfold is considered to be significant inhibition.

* Trypsin is added to cleave the HA_0 to HA_1 and HA_2 proteolytically, and enhance infectivity of influenza in cell culture. Addition of FCS would abrogate the effect of the protease and hence is not added.

A major disadvantage of this technique is that once the endpoint is detected the assay is over, unlike an experiment which includes sampling the supernatant, which can be done repeatedly, limited only by the quantity of supernatant present, thus allowing the viral infection to be monitored over a time course. The optimum time at which to end the experiment can be determined by careful monitoring of the viral-induced CPE, using an inverted microscope. However, this increases the labour involved, and therefore reduces the applicability of this technique to automation.

Protocol 4. *In vitro* testing of multiple compounds against influenza in multiwell plates

Equipment and reagents

- 48- or 96-well tissue culture plates (e.g. Falcon, Nunc, etc.)
- Minimum Essential Medium containing 50 IU ml^{-1} penicillin, 50 μg ml^{-1} streptomycin, and 2 mM L-glutamine (Sigma, Life Technologies, ICN Flow, etc.)
- Gilson pipette (100 μl) and sterile tips
- Phosphate-buffered saline (PBS)
- 5 ml and 10 ml sterile plastic pipettes (Phillip Harris)
- Test compound at predetermined concentrations
- 0.05% TPCK trypsin (Worthington)
- Previously titrated influenza virus preparation
- Amantadine (test control substance) (Noveargis Pharmaceuticals UK Ltd)

Method

1. Culture MDCK cells according to standard procedures to produce confluent monolayers in 48- or 96-well plates.
2. Dilute the compound in tissue culture medium to the required concentration, and add to the appropriate wells on the plate using a Gilson pipette.
3. Incubate the cells for 1 h at 37°C in a CO_2 incubator, to allow the compound and cells to equilibrate.
4. Using a Gilson pipette, add 50 μl of tissue-culture medium containing 100 $TCID_{50}$ of influenza virus to the appropriate well on the plate. The plate layout should include cells that are not challenged with virus but are exposed to the highest concentration of the compound to be tested, as a control for potential compound cell toxicity.
5. Incubate the plates for up to 5 days, and monitor CPE closely using an inverted microscope.
6. Titrate the virus, and determine the endpoint by either HA assay, staining the remaining monolayer with crystal violet, or by using the MTT system described below.
7. The ED_{50} (the dose required to reduce the titre of the virus by 50%) and/or the ED_{90} (the dose of compound required to reduce the titre of the virus by 90%, i.e. one log reduction) can then be calculated.
8. In addition the Therapeutic Index of the test compound can be calculated using the following formula:

   Therapeutic Index (TI) = CyD_{50} / ED_{50}

   where CyD_{50} is the maximum compound concentration that causes cytotoxic effects in 50% of the cultured cells and ED_{50} is the minimum compound concentration that is effective to inhibit virus induced cytopathic changes by 50%.

Another method to detect the endpoint uses a modification of the MTT assay technique, which is described below for testing anti-herpes compounds (*Protocol 5*). This can be further automated by using standardized plate layouts, automatic transfer of data from the ELISA reader to a computer system, and then a spreadsheet programme to calculate the ED_{50} of the test compounds. Internal standards on the plate, and suitable controls with known performance criteria, will act to validate each individual plate. The software can be programmed to highlight automatically those plates where the controls have failed. Specialist software is not required, and standard spreadsheet packages such as Microsoft Excel® and Lotus 123® are more than capable of performing this role.

4.2 *In vitro* testing of compounds for antiviral effect against herpes simplex

In our laboratories, we prefer a colorimetric MTT assay system for the detection of the cytopathic effect of herpes simplex type 1. This is an established method to evaluate the effect of a compound against a cytopathic virus, and can also be used for anti-HIV screening (15). MTT is 3-(4,5-dimethylthiazol-2-yl)-2,5-di-phenyl-tetrazolium bromide, which is metabolized by healthy cells to an insoluble coloured formazan product (16) as shown in *Figure 5*. The level of metabolism that occurs in the individual wells of a 96-well plate will be dependent on the number of healthy viable cells present, which itself will be inversely proportional to the level of CPE and cell destruction caused by the virus infection. If the test compound inhibits viral replication, then fewer cells will be lysed by viral replication, and higher levels of the coloured metabolite will be present. The MTT-formazan product is insoluble, and it is necessary to disrupt the cell membrane and solubilize the product using Triton X-100 and isopropanol (*Protocol 5*).

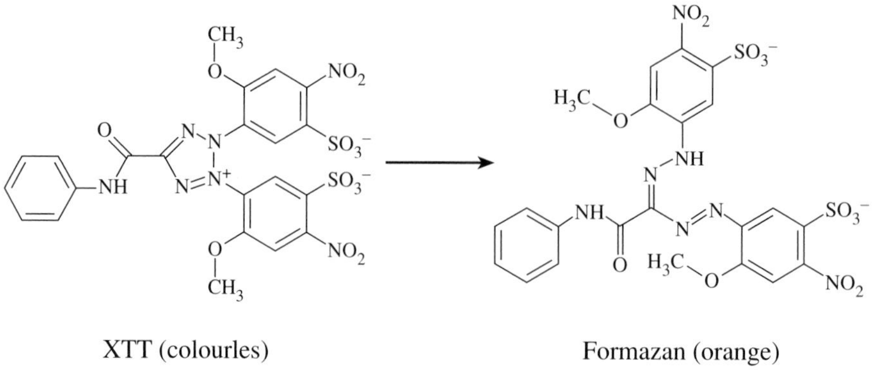

Figure 5. Basis of the XTT tetrazolium to XTT formazan assay.

Protocol 5. *In vitro* testing of potential anti-herpes compounds using the MTT antiviral assay system

Equipment and reagents

- 48- or 96-well tissue culture plates (e.g. Falcon, Nunc, etc.)
- Multichannel pipette (Anachem) or digital multidispensing pipetter (e.g. Jensen Seal-pipette)
- Phosphate-buffered saline (PBS)
- 0.05% TPCK trypsin (Worthington) in PBS
- Minimum Essential Medium containing 50 IU ml^{-1} penicillin, 50 μg ml^{-1} streptomycin, and 2 mM L-glutamine (Sigma, Life Technologies, ICN Flow, etc.)
- Suitable virus
- MTT (Sigma) 5 mg ml^{-1} in PBS
- 10% SDS + 0.01 N HCl

Method

1. Cultivate BHK cells and HSV-1 according to standard procedures.
2. Add 100 μl of viable BHK-21 cells, at a concentration of 2×10^6 cells per ml^{-1} as determined by trypan blue exclusion, to each well of a 96-well flat-bottomed tissue culture plate, using a multi-channel pipette and sterile tips.
3. Incubate the cells for 48 h at 36.5°C in a 5% CO_2 incubator or until confluent.
4. Wash the cells by the addition and aspiration of 200 μl of sterile PBS, using a multi-channel pipette and sterile tips.
5. Add 100 μl of the test compound or acyclovir, diluted to known concentration in tissue culture medium, using a multichannel pipette.
6. Incubate the cells for 1 h at 36.5°C in a 5% CO_2 incubator.
7. Add 10 μl of 100 $TCID_{50}$ of virus, as determined by the Karber method (*Protocol 6*), or medium as control, to all wells containing the test compound, using a multichannel pipette
8. A single 96-well plate is used for infected cells, and a separate plate for non-infected cells, to prevent possible contamination. Each assay is performed in triplicate.
9. Incubate the cells for 3–5 days at 36.5°C in a 5% CO_2 incubator.
10. Monitor the level of viral-induced CPE, using an inverted microscope.
11. Add 10 μl of 5 mg ml^{-1} MTT in PBS to each well, using a multichannel pipette
12. Incubate the cells for 4 h at 36.5°C in a 5% CO_2 incubator.
13. Add 100 μl of isopropanol in distilled water, using a multichannel pipette
14. Incubate the cells for 18 h at 36.5°C in a 5% CO_2 incubator.
15. Record the absorbance of each well using an ELISA reader, at a wavelength of 540 nm.

Protocol 6. The Karber method for the determination of $TCID_{50}$

Equipment

- Scientific calculator

Method

This is a simple mathematical method for the calculation of the $TCID_{50}$ of a certain virus, and it uses of the following equation:

$$-\text{Log } TCID_{50} = L - d\,(S - 0.5)$$

where:
L = log dilution of the strongest concentration
d = difference between successive log dilutions
S = the sum of proportional positive tests

Examples of how to use the Karber method to calculate $TCID_{50}$

Example 1:

Virus	+	+	+	−	−	−	−	−	−	−	−	−
	+	+	−	−	−	−	−	−	−	−	−	−
	+	+	−	−	−	−	−	−	−	−	−	−
	+	+	−	−	−	−	−	−	−	−	−	−
Control												

+ = positive for virus growth
− = negative for virus growth

Explanation:

If the first row is a tenfold dilution, then the log of this is 1.
If the successive dilutions are also tenfold, then the log of this is also 1. The sum of the proportional positive responses is 2.25, i.e. the first two rows are positive and 1 out of the 4 wells in the third row is positive.
If these values are put into the equation we get:

$$-\text{Log } TCID_{50} = 1 - 1\,(2.25 - 0.5)$$
$$\text{Log } TCID_{50} = 1.75$$

Example 2:

Virus	+	+	+	+	+	−	−	−	−	−	−	−
	+	+	+	+	+	−	−	−	−	−	−	−
	+	+	+	+	−	−	−	−	−	−	−	−
	+	+	+	+	−	−	−	−	−	−	−	−
Control												

+ = positive for viral growth
− = negative for viral growth

Explanation:
Again the first row is a tenfold dilution, and the log of this is 1.
Again the successive dilutions are also tenfold, and the log of this is again 1.
The sum of the proportional positive responses is 4.5, i.e. the first four rows are positive and 2 out of the 4 wells in the fifth row are positive.
If these values are put into the equation we get:

$$-\text{Log TCID}_{50} = 1 - 1\,(4.5 - 0.5)$$
$$\text{Log TCID}_{50} = 4$$

4.3 *In vitro* testing of compounds for antiviral effect against HIV-1

Since the discovery of HIV-1 in 1984, the virus replication cycle has been extensively studied. Various compounds have been shown to demonstrate different efficacies against HIV, depending on both the cell line used and the strain of virus chosen for the challenge. In our laboratories, initial screening for activity against HIV is conducted in both C8166 cells and PBMCs. Initially, we test a compound against the two common laboratory strains of clade B, namely HIV-1_{IIIB} and HIV-1_{MN}, and if antiviral activity is found the compound is then tested against two other recent clinical virus isolates from clades C and E, which are more typical of the 'wild-type' virus, and also against HIV-2.

Compounds with virucidal effects are of particular interest for HIV, with potential uses as surface disinfectants, or to prevent the transmission of the virus during sexual intercourse by their incorporation in pessaries or lubricants. Below are the protocols commonly used in our laboratories to test for virucidal or virustatic activity of anti-HIV compounds. *Protocol 7* describes a simple cell toxicity assay based on trypan blue exclusion whilst *Protocol 8* details titration of HIV to determine 100 TCID_{50} for subsequent virucidal (*Protocol 9*) and virustatic (*Protocol 10*) tests.

Protocol 7. Determination of cellular toxicity of the compounds for C8166 cells

Equipment and reagents

- 25cm^2 cell culture flasks (e.g. Corning, Falcon, Costar, etc.)
- C8166 cells
- Haemocytometer (e.g. Sigma)
- Gilson pipette
- RPMI 1614 medium containing fetal calf serum, 2 mM L-glutamine, 100 units ml^{-1} penicillin, 100 μg ml^{-1} streptomycin (Sigma, Life Technologies, ICN Flow, etc.)

Method

1. Cultivate C8166 cells according to standard procedures.
2. Pellet the C8166 cells (approx. 5×10^5 cells per ml), and resuspend in four concentrations of the test compound.

Protocol 7. *Continued*

3. Use cells resuspended with growth medium alone as negative controls.
4. Add the cell/drug mixtures (1 ml) to tissue culture flasks.
5. Test the viability of the cells by trypan blue exclusion on days 3, 5, and 7.
6. Mix 90 μl from each well with 10 μl of trypan blue, and transfer onto a haemocytometer for counting.

Protocol 8. Titration of HIV virus in 96-well plates

Equipment and reagents

- 96-well tissue culture plates (e.g. Nunc)
- C8166 cells
- Virus stock
- Gilson pipette
- RPMI 1614 medium containing fetal calf serum, 2 mM L-glutamine, 100 units ml^{-1} penicillin, 100 μg ml^{-1} streptomycin (Sigma, Life Technologies, ICN Flow, etc.)

Method

1. Plate out 180 μl of lymphoblastoid C8166 cells into 96-well plates at a concentration of 5×10^5 cells per ml, in six replicates.
2. Add 20 μl of undiluted HIV-1_{IIIB} stock to the first row.
3. Mix the virus-cell mixture, and then transfer 20 μl into the next series of wells etc. to give a tenfold dilution series from 10^{-1} to 10^{-7}, using a Gilson pipette and sterile tips.
4. The last row is used as a negative control without virus.
5. Incubate the cells at 37 °C in 5% CO_2 for seven days.
6. The production of p24 antigen or syncytia is used to determine the endpoint.
7. Calculate the $TCID_{50}$ virus titre using the Karber method (*Protocol 6*).
8. It is important to conduct virus titrations before and also in parallel with the assay, to ensure optimal accuracy of the virus input dose.

Protocol 9. Determination of the virucidal effects of a compound against HIV-1

Equipment and reagents

- Virus preparation of known titre (*Protocol 8*)
- Sodium deoxycholate (100 μg ml^{-1}) as a positive control drug
- 96-well tissue culture plates (e.g. Nunc)
- C8166 cells
- RPMI 1614 containing fetal calf serum, 2 mM L-glutamine, 100 units ml^{-1} penicillin, 100 μg ml^{-1} streptomycin (Sigma, Life Technologies, ICN Flow, etc.)

Method

1. Incubate four different concentrations of each compound with both a high titre of HIV-1_{MN} (10^6 $TCID_{50}$) and low titre of HIV-1_{MN} (10^3 $TCID_{50}$) at a ratio of 1:1 (1 ml total) at 37°C for 1 h.
2. Use sodium deoxycholate (100 μg ml^{-1}) as the positive control for this assay.
3. Meanwhile, distribute 180 μl per well of C8166 cells in growth medium, at a concentration of 2.5×10^5 per ml, into a 96-well plate.
4. Add 20 μl from each concentration of the virus for each compound mixture into the top wells (quadruplicate for each dilution), and then 20 μl from the top wells is transferred into the next series of wells. Repeat this to give a tenfold dilution series from 10^{-1} to 10^{-7}.
5. Use the last row as a negative control without the compound-virus mixture, and incubate further at 37 °C for 7 days.
6. Score the virus-induced syncytia on days 3, 5, and 7.

Protocol 10. Determination of the virustatic effects of the compounds against HIV-1

Equipment and reagents

- Virus preparation of known titre (*Protocol 8*)
- Zidovudine (AZT)
- 96-well tissue culture plates (e.g. Nunc)
- C8166 cells
- RPMI 1614 containing fetal calf serum, 2 mM L-glutamine, 100units ml^{-1} penicillin, 100 μg ml^{-1} streptomycin (Sigma, Life Technologies, ICN Flow, etc.)

Method

1. Four different compound concentrations are used that have previously been determined not to exert a cytotoxic effect on the cell line.
2. The positive control is the nucleoside analogue AZT at concentrations of 10 μM, 5 μM, and 1 μM.
3. Incubate four different concentrations of each compound with C8166 cells in universal tubes at 37 °C for 30 min.
4. After incubation, add high (10^6) and low (10^3) titres of HIV-1 to the appropriate universal tube, except the control tubes, and incubate for a further 3 h at 37 °C.
5. Wash the cells in the universal tubes twice with growth medium by low speed centrifugation, to eliminate the residual unbound virus.
6. Re-suspend the cell pellets from each tube in 2 ml of growth medium, and plate out in duplicate in a 48-well plate (1 ml per well).

Protocol 10. *Continued*

7. Add 100 μl of 10× compound concentration to the appropriate wells.
8. Samples are taken at days 3, 5, and 7, and the viral p24 antigen levels determined by using a p24 ELISA.

4.4 Medium to high throughput screening for potential antiviral compounds effective against HIV-1

The above methods are suitable for testing a relatively small number of compounds against HIV, but the methodology would be difficult to automate and scale up for medium to high throughput screening. There is therefore a need for a more rapid method to conduct initial screening of compounds. Those that are found to effective can then be tested further, as described above. A suitable system is a variation on the MTT-formazan based assay described earlier for testing compounds against HSV-1. The compound XTT (2,3-bis[2-methoxy-4-nitro-5-sulfophenyl]-5-[(phenylamino)carbonyl]-2H-tetrazolium hydroxide) is used and, like MTT, it is metabolized by healthy cells to form an XTT-formazan product (*Protocol 11*). However, this product is soluble, and therefore it is not necessary to disrupt the cells and solubilize the insoluble product as with MTT-formazan, thus further reducing the work involved. Assay systems using similar reagents are also commercially available (e.g. Promega CellTiter 96™; Boehringer Mannheim Cell Proliferation Kits).

Protocol 11. Primary screening of antiviral compounds using the XTT formazan method

Equipment and reagents

- Virus preparation of known titre (*Protocol 8*)
- Zidovudine (AZT) as a positive control
- 96-well tissue culture plates (e.g. Nunc)
- C8166 cells
- XTT (Sigma)
- RPMI 1614 containing fetal calf serum, 2 mM L-glutamine, 100units ml^{-1} penicillin, 100 μg ml^{-1} streptomycin (Sigma, Life Technologies, ICN Flow, etc.)
- N-methylphenazonium methosulfate (Sigma)

Method

1. Make six serial fivefold dilutions of each compound in triplicate in 50 μl of medium in a 96-well tissue culture plate, using a multichannel pipette.
2. Add 25 μl of C8166 cells at a concentration of 1.6×10^5 per ml to each well.
3. Add 25 μl of a known infectious titre of HIV-1_{IIIB} or HIV-1_{MN} supernatant that would kill approximately 70–90% infected cells within 5 days.

4. Include untreated wells as controls for both infected and uninfected cells.
5. Incubate the plates for 5 days at 37 °C and 5% CO_2.
6. Examine the cell cultures after 3 days for CPE, using an inverted microscope.
7. On day 5 add 25 μl of a mixture of XTT, 2mg ml^{-1}, and 0.02 mM N-methylphenazonium methosulfate to infected and uninfected cultures.
8. Incubate plates for 4 h at 37 °C.
9. Agitate the plates for 1 h using a plate shaker.
10. Read the optical densities at a test wavelength of 450 nm, and at a second wavelength of 650 nm to control for non-specific absorbance.

5. Animal model infections

A clinical antiviral effect of a compound in an animal model represents an important stage in the development of a new antiviral. Such studies are normally a prerequisite for clinical trials. However, only highly active compounds will be tested *in vivo*, as the experimental work is relatively expensive. Modern experimental animal units are carefully inspected, and all staff are trained in animal care and experimental techniques. In the UK, every experiment is detailed in a Project License, and can only be carried out by trained scientists who, in addition, have a personal license from the Home Office. It is important to appreciate that an effect of the drug on both clinical and virological aspects of the animal model infection will be sought. Often a reduction in virus infectivity titre of only tenfold will have a significant effect on the clinical parameters in the model. Important experimental parameters are the infectious titre of the input challenge virus, and the clinical scoring technique. Rarely would an antiviral compound completely prevent an infection in an animal model, unless the experiments are adjusted to encompass low infectivity titres of challenge virus, together with a high drug dose. Obviously the drug should be present in the target organ for viral replication, namely lung and respiratory tract for influenza virus, skin for herpes virus, and spleen and lymph nodes for Rauscher Leukaemia virus (a model for retrovirus infection), at concentrations predicted from the *in vitro* data to inhibit virus replication. Animal experiments can also on occasion give the first evidence of the emergence of drug-resistant virus (17).

5.1 Influenza virus animal model systems

Classically, the first isolation of influenza virus in 1933 used the ferret (18). The animals are easily infected with 0.5 ml of virus dropped into the nostrils under light anaesthesia with a mixture of oxygen and halothane. Within 48 hours the animals become ill, listless, and have a raised temperature. The infection is not lethal unless the virus is passaged in the ferret to increase virulence deliberately. In a pre-titration experiment, varying tenfold dilutions of influenza virus, either the clinical isolate itself, or egg-grown or MDCK-cell grown passages, are inoculated intranasally using 3-4 ferrets per dilution. Commonly, 100 EID_{50} of virus will infect a ferret and initiate virus replication in the upper respiratory tract, but will cause no clinical symptoms. In contrast, infection with 1000 EID_{50} will normally induce a temperature response on day three, and also a rapid increase in the number of immune cells.

The animals are housed 'free range' in groups in large floor pens with plastic drain pipes and boxes for exercise. The animals are checked daily under light anaesthesia for five consecutive days for temperature rise. A digital thermometer is inserted anally. At the same time, whilst the animal is lightly anaesthetized, the nasal cavity is washed out using 5–10 ml of saline. The animal is induced to sneeze, and the washing collected in a large plastic container, into which the ferret's head is briefly inserted. Scientific personnel normally wear gowns, gloves, and face masks, because influenza has been transmitted from the ferret to a scientist, although this must be a very rare event. If clinical virus isolates are being investigated, which have had relatively few passages *in vitro,* the scientist should be vaccinated with the current influenza vaccine. In addition, because of the risk of infection, no 'at risk' person (elderly, diabetic, asthmatic, or a person with chronic heart or kidney disease) should participate in ferret experiments. However many experiments utilize laboratory strains of influenza which have been extensively passaged in the laboratory, and these are extremely unlikely to be infective for laboratory or animal house personnel.

The nasal washings are frozen at $-70\,^{\circ}C$ in aliquots, and subsequently titrated on MDCK cells for infectious virus (*Protocol 1*). A parallel nasal wash sample is immediately processed to quantify the number of immune cells, using a haemocytometer and the trypan blue exclusion test. Lung lavage may also be performed at the end of the experiment on day six, to quantify both immune cells and virus. In addition, the recovery of influenza virus from the lung tissue can be titrated on MDCK cells. Infected ferrets with clinical symptoms also lose weight, and this is an extra parameter to quantify antiviral effects. Predetermined concentrations of the antiviral are administered orally twice daily, diluted in liquid cat food. For the antiviral test, approximately 1000 EID_{50} of influenza virus is used in the challenge dose. Challenge with higher doses of influenza may result in 'overchallenge' and a reduction in the sensitivity of this assay system.

Another commonly used model for influenza is a lethal pneumonia infection in laboratory mice. This system has several benefits when compared to the ferret model, not the least of which is easier handling and reduced cost. However, influenza viruses, whether clinical isolates or laboratory passaged strains, will not readily infect the mouse and produce disease unless first adapted to the new host system by repeated passage. A representative influenza A (H_1N_1), (H_3N_2), and an influenza B strain are therefore first adapted by rapid intranasal passage of 0.1 ml virus at 3–4 day intervals in BALB/c mice using light anaesthesia of oxygen and halothane. Lung adaptation of the virus is variable, and may or may not be achieved after 10–20 passages in the lung. Transmission of a mouse-adapted influenza virus to a human has never been described, and therefore experiments may be carried out safely in a class II laboratory environment (*Protocol 13*).

Protocol 12. Antiviral testing of compounds against influenza, using the ferret animal model

Equipment and reagents

- Liquivite® (Creg Vetfoods)
- Electronic tags for animal identification (AVID Labtrack)
- Titrated influenza virus preparation
- Halothane and oxygen for anaesthesia

Method

1. Anaesthetize three groups of five ferrets with halothane, and infect intranasally with approximately 1000 EID_{50} of influenza virus in 0.5 ml of saline.
2. Prepare the test compound at two different concentrations in liquid cat food (Liquivite®)
3. Dose five animals with the vehicle (Liquivite®), five with low dose, and five with high dose of compound, twice daily at 12 h intervals.
4. Commence dosing 2 h before infection, and continue for 6 days.
5. Obtain nasal wash samples twice daily from each animal, and record the time relative to the last dose.
6. Obtain nasal wash samples taken in turn from an animal in the vehicle group, one in the low-dose, and one in the high-dose group, before returning to the vehicle group. In this way, samples are obtained at a variety of times post-dose for each group.
7. Transfer the nasal wash to a 20 ml sterile universal tube, and store on ice.
8. Measure the number of inflammatory cells, using a haemocytometer.
9. Measure the influenza virus titre of the nasal wash on MDCK cells.
10. On day 6 of the experiment, animals are given a final dose of vehicle or compound, then terminal lung lavage and blood samples are obtained via cardiac puncture at recorded times post-dose (ideally 1, 2, 4, 6, and 8 h post-dose if the drug itself is to be quantified by HPLC, for example).

Protocol 12. *Continued*

11. Lungs can be removed and retained for histology and/or virus titre determination.
12. Collect blood samples into EDTA tubes and hold on ice.
13. Spin at 4 °C to remove red blood cells. Store plasma at −20 °C.
14. Lavage samples can be treated in the same way as nasal-wash samples.
15. Ideally the study should be repeated for recent examples of influenza A (H_1N_1), influenza A (H_3N_2), and influenza B viruses.

Protocol 13. Adaptation of an influenza virus for the mouse host

Equipment and reagents

- Groups of ten BALB/c mice for each passage
- Sterile surgical scissors
- Sterile surgical forceps
- 20 ml sterile universal tubes
- Sterile pestle and mortar
- Sterile sand
- Minimum Essential Medium (MEM; Sigma, Life Technologies, ICN Flow, etc.)
- Class II safety cabinet
- Benchtop centrifuge
- Pasteur pipettes

Method

1. Infect mice with 0.1 ml of influenza virus (approximately 10 000 EID_{50}) under light anaesthesia with halothane.
2. Kill the mice after 3–4 days, by asphyxiation with CO_2 or by cervical dislocation.
3. Expose the thoracic cavity by incision with surgical scissors, and remove the lungs.
4. Place the lungs in individual 20 ml sterile plastic universals.
5. Mix the lungs individually with approximately 1 g of sterile sand, and grind for 2 min with a mortar and pestle at approximately 4 °C (i.e. supported on a bucket of ice).
6. Add 5 ml of MEM to the suspension, and centrifuge at 1500 r.p.m. for 10 min to deposit large tissue fragments.
7. Remove the supernatant fluid containing virus, using a sterile Pasteur pipette.
8. Aliquot the supernatant, and store all but one at −70 °C.
9. Use the remaining aliquot to infect a new group of ten mice.
10. Repeat the above procedure for each passage.
11. After approximately three passages, titrate a second aliquot of lung tissue supernatant for the presence of influenza virus, using MDCK cells in parallel.

12. Observe the mice for clinical symptoms and mortality.
13. Adaptation of the virus will often occur after 10–20 passages, and such a virus will produce clinical symptoms in mice, and detectable influenza virus in the lungs.
14. Further passage of the virus in mice will cause enhanced adaptation such that the virus will produce fatal pneumonia after 5–10 days in the mouse model.

The mouse influenza model can be used in two ways (*Protocol 14*). It can be used as a lethal pneumonia system, whereby the efficacy of the test compound is determined by its ability to prevent mortality. An alternative mouse influenza model allows a quantification of the virus present in the mouse lung before mortality occurs. In this system, the test compound is injected into the peritoneal cavity of the mouse (i.p.) in groups of ten. The mice are then maintained for one to two hours, and then infected with a pre-determined infectious dose of influenza virus of an appropriate sub-type. The number of animals that die of lethal pneumonia in both the treated and control groups is monitored over a 12-day period. A parallel experiment can be conducted, in which the animals are killed at day 4–5 and the lungs removed. The titre of influenza virus is then obtained by titration of a suspension of lung tissue on MDCK cells. The compounds ribavirin or amantadine are used as positive control drugs, as they have known efficacy against influenza viruses.

Protocol 14. Antiviral testing of compounds against influenza using the mouse model

Equipment and reagents

- Groups of ten BALB/c mice
- Sterile surgical scissors
- Sterile surgical forceps
- 20 ml sterile universal tubes
- Sterile pestle and mortar
- Sterile sand
- Minimum essential Medium (MEM; Sigma, Life Technologies, ICN Flow, etc.)
- Class II safety cabinet
- Benchtop centrifuge
- Pasteur pipettes

A. *Methods*

1. Electronically tag all mice on day 0.
2. Divide 60 female BALB/c mice, randomized by litter, into six groups, each containing 10 mice.
3. Group 1 to 5 are administered the test compound at varying concentrations and time periods as appropriate.
4. Challenge each sub-group of ten mice intranasally with the mouse-adapted virus of the appropriate influenza subtype under light anaesthesia using halothane.

Protocol 14. *Continued*

5. Sacrifice the mice at day 5 by asphyxiation with CO_2.
6. Expose the thoracic cavity by incision with surgical scissors and remove the lungs.
7. Place the lungs in individual 20 ml sterile plastic universals.
8. Mix the lungs individually with approximately 1 g of sterile sand, and grind for 2 min with a mortar and pestle at 4°C (supported on a bucket of ice).
9. Add 5 ml of MEM to the suspension, and centrifuge at 1500 r.p.m. for 10 min to deposit large fragments.
10. Remove the virus-containing supernatant fluid, using a sterile Pasteur pipette.
11. Store the supernatant in suitably labelled sterile tubes at −70°C.

B. *Titration of influenza virus from lungs*

1. Prepare confluent monolayers of MDCK cells in 48-well plates.
2. Remove the supernatant, and wash the cells twice with 200 μl of sterile PBS.
3. Add 460 μl of DMEM to each well on the plate. No FCS is added.
4. Add 2.5 μg ml^{-1} of TPCK trypsin.
5. Keep the lung homogenate on ice throughout this procedure.
6. Add 40 μl of lung homogenate to the first well of the prepared plate.
7. Conduct the assay in duplicate.
8. Dilute the lung homogenate across the plate in tenfold steps to give a dilution range of 10^{-1} to 10^{-7}.
9. Incorporate a negative control of medium only, and a positive control of a stock virus.
10. Incubate the plates in a 5% CO_2 incubator at 37°C for 3–5 days.
11. Remove the supernatant from each well, and titrate the presence of influenza virus by haemagglutination assay (*Protocol 2*).
12. Calculate the mean viral titre and standard deviation from each group, and the compound-treated group compared to the non-compound-treated group.

5.2 Mouse retrovirus model

The Rauscher leukaemia virus (RLV) mouse model system is an accepted method of determining the efficacy of compounds against retroviruses, and therefore their potential efficacy against the human retrovirus HIV (19). RLV infects the spleen, and causes 2–4 fold enlargement (*Protocol 15*). This small-animal model infection can be used to screen potential anti-HIV drugs before testing in the far more expensive animal models using primates (SIV) (20, 21). This RLV assay system does suffer from the simple fact that it is using a surrogate virus rather than the intended target virus. There is no perfect animal model for HIV since the only animal infected, apart from humans, is the chimpanzee. These are protected animals and, moreover, the virus does not produce any symptoms of disease. Because of these acknowledged difficulties, many anti-HIV drugs move directly from *in vitro* cell culture and viral enzyme testing into phase I clinical trials in human volunteers.

The RLV is stored at $-70\,°C$ as a 10% spleen suspension supernatant. A preliminary titration of the virus stock in Balb/C mice gives information on the MID_{50} of the aliquotted virus pool. Commonly 1/5 to 1/50 or 1/500 dilutions of the stock virus induce spleen enlargement. In the test experiment, the first dose is given one hour before challenge with the virus, and thereafter the drug is administered twice daily and the virus given intraperitoneally (0.1 ml). The anti-HIV compound zidovudine (AZT) is used as a positive control. Infection of the mouse spleen leads to rapid enlargement of this organ, and the animals are killed after 8–14 days, when up to a fivefold increase in spleen size is noted. Spleens from groups of mice are weighed and compared to the size of these in compound-treated but uninfected mice, in virus-infected mice, and in normal mice. The ability of the test compound to prevent an increase in spleen size is used to determine its efficacy as an antiviral compound.

Protocol 15. Rauscher leukaemia virus mouse model system

Equipment and reagents

- Sixteen female BALB/c mice
- Sterile saline
- Rauscher leukaemia virus
- Test compounds
- Zidovudine (AZT) (positive control drug)
- Sterile surgical scissors
- Sterile surgical forceps
- 20 ml sterile universal tubes
- Sterile pestle and mortar
- Sterile sand
- Minimum Essential Medium (MEM; Sigma, Life Technologies, ICN Flow, etc.)
- Class II safety cabinet
- Bench centrifuge
- Pasteur pipettes

A. *Toxicity testing of the compounds* in vivo

1. Electronically tag all mice before day 0, and check the tag on day 0.
2. Divide sixteen female BALB/c mice, randomized by litter, into four groups.
3. On day 0 inject each group intraperitoneally with either the test compound at differing concentrations, or 200 μl of saline.
4. Inject the mice twice daily for 10 days with the appropriate concentrations of compound or saline.
5. Monitor the mice over the 10 days to assess whether the compounds are having a toxic effect on the mice.

Protocol 15. *Continued*

6. Record the weight of every mouse every two or three days.
7. Sacrifice the mice by cervical dislocation.
8. Based on the observations in this experiment, select the drug concentrations for the antiviral assay.

B. *Titration of RLV* in vivo

1. Electronically tag all mice before day 0, and the check the tag on day 0.
2. Divide twenty female Balb/c mice, randomized by litter, into five groups, each containing four mice.
3. Inject each group intraperitoneally with a dilution of the virus stock (either 1/5, 1/50, 1/500, or 1/5000).
4. Inject the final control group with 200 μl of saline.
5. Monitor the mice over 14 days for spleen enlargement.
6. Weigh the mice every two or three days.
7. On day 6 sacrifice two mice from each group, remove their spleens, and store at −70 °C.
8. On day 14 sacrifice the remaining mice by cervical dislocation, remove their spleens, and store at −70 °C.
9. Weigh the spleens to assess the extent of the spleen enlargement.
10. Use this data to calculate the MID_{50}.

C. *Determination of the antiviral effect of the test compounds*

1. Electronically tag all the mice before day 0, and check the tag on day 0.
2. Use two concentrations of virus of high and low infectious titre, which are known to cause an appropriate level of spleen enlargement.
3. On day 0 divide the mice into groups of twelve, and administer either the test compound, zidovudine (AZT), or saline.
4. Subdivide each group, and challenge half with a low dose of virus and the other half with a high dose of virus, as determined in the earlier experiment.
5. Over the following 14 days inject the mice as appropriate with the test compound.
6. Monitor the mice for splenomegaly over the experimental period.
7. Sacrifice the mice on day 14 by cervical dislocation, then remove and weigh the spleens.
8. Calculate the mean weight increase and the standard deviation. Compare the treated and non-treated groups.

5.3 Herpes simplex guinea pig model

Two commonly used model infections are the guinea pig skin model (HSV-1) (*Protocol 16*), and vaginal infection of the guinea pig (HSV-2) (22–27) (*Protocol 17*). In the skin model system, a small area (25 cm^2) of the flank of the animal is shaved free of hair, and lightly scratched in a criss-cross manner in nine previously marked areas. A range of concentrations of herpes virus may be added dropwise to the skin to initiate infection. Antiviral compounds are usually tested as a topical preparation, with matched controls, and the antiviral effect is scored by decrease in the healing time, or, if administered very early in infection, by prevention of the herpes skin lesion. The HSV-2 vaginal model is similar, and the protocols for both techniques are given below. Acyclovir is commonly used as a control compound.

Protocol 16. Guinea pig skin lesion model system for HSV-1 infection

Equipment and reagents

- Guinea pigs
- Herpes simplex virus type 1
- Compounds to be tested as topical preparations
- Acyclovir as a topical preparation (Zovirax™ 5% cream)
- Sterile surgical scalpels

Method

1. Shave a small area (25 cm^2) of the flank, and lightly scratch in a criss-cross manner in nine areas.
2. Add a range of concentrations of herpes virus dropwise to the skin in a total volume of 100 μl.
3. The HSV-1 infects the epithelial cells, and forms a small vesicle that gradually heals after crusting.
4. Administer the test compounds topically in a suitable cream base.
5. Apply acyclovir 5% topical cream (Zovirax™) as a positive control.
6. Apply the cream base used with the test compound as a negative control.
7. Apply carefully weighed quantities of the creams five times daily.
8. Observe the animals daily, and record the antiviral effect by scoring the decrease in the healing time, or, if the test compounds are administered very early in infection, by prevention of the herpes skin lesion.
9. Quantify the virus by taking swabs from the lesion and titrating on BHK-21 cells.

Protocol 17. Guinea pig genital model system for herpes simplex virus type 2 infection

Equipment and reagents

- Guinea pigs
- Herpes simplex virus type 2
- Compounds to be tested
- Acyclovir as a topical preparation (Zovirax 5% cream)
- Sterile surgical scalpels

Method

1. Using a cotton swab impregnated with a range of concentrations of herpes virus, infect guinea pigs intravaginally in groups of ten.
2. HSV-2 infects the epithelial cells, causes erythema, and a crop of herpes lesions.
3. Administer the antiviral compounds topically in a cream base, orally up to five times a day mixed in the animals' food supply, or by injection intraperitoneally.
4. Use acyclovir as a positive control.
5. The herpes lesions on the external genitalia are quantified daily.
6. Isolate virus by taking swabs from the lesion, and titrate on BHK-21 cells.

At varying times after healing, the previous genital infection may recur, and so a further analysis may be performed of the ability of an antiviral to prevent recurrent herpes infection. Human herpesviruses are also easily adapted to infect mice, either as a skin lesion, or as a lethal infection of the brain following intracerebral inoculation.

6. Genetic analysis of virus resistance to antiviral drugs

It is now accepted that, should drug-resistant mutants of a virus not be isolated, then the drug is not exerting an antiviral effect. In the case of DNA genome viruses, where the viral DNA polymerase is a accurate transcriptase and, moreover, has editing and correction functions, mutation rates are 10 000 times less frequent than with RNA genome viruses. Therefore, in general terms, resistance is not a clinical problem with herpesviruses in the normal non-immunosuppressed patient. But even with high fidelity transcription, the strong selective pressure exerted by an antiviral drug will allow the emergence of drug-resistant viruses. Fortunately, there is some evidence that such viruses may be less infectious, and at least attenuated, and hence have fewer opportunities to spread. But drug-resistant herpes viruses do occur (28) and are a problem in immunosuppressed persons, especially those with HIV infection

(29). In contrast, most RNA viruses are highly polymorphic and exist as a quasispecies, and it highly unlikely that, within any two persons or even within a single person, a single genome sequence exists. It is not, therefore, surprising that drug-resistant mutants have been described for every compound discovered to date. In fact, such is the mutation rate of quasispecies RNA viruses such as HIV and influenza that drug-resistant mutants are thought to predate the discovery of an inhibitor.

The rapid molecular techniques of PCR and nucleotide sequencing can now be used to probe the nucleotide sequences of important viral genes, such as HIV reverse transcriptase or protease, or influenza neuraminidase, to detect changes which are known to be associated with drug resistance. The use of indirect methods of DNA genotyping (such as DNA probes) are not effective at detecting the presence of compound-resistant mutants, as they require precise knowledge of the DNA sequence. The only effective option is to sequence the region where expected mutations would occur that would confer compound resistance. In addition, close monitoring of a patient's plasma viral load can be used as an early indicator of emerging drug resistance. Alternatively, an individual may already be infected with a resistant virus, and the anticipated reduction in viral titre will not be observed during treatments.

Commercial kits exist for the sequencing of the protease and reverse transcriptase regions of HIV, such as the TrueGene Kit by Visible Genetic Inc. (`http://www.visgen.com`) which uses proprietary technology known as CLIP™ (*Protocol 18*). If drug-resistant mutants are identified, then a new antiviral drug can be substituted, or better still a combination of drugs, each acting on a different gene product utilized, which may reduce the likelihood of drug-resistant mutants emerging.

Protocol 18. Identification of HIV compound-resistant mutants using the TrueGene™ system

Equipment and reagents

- TrueGene Kit (Visible Genetic Inc.)
- Automated DNA sequencer

Method

1. Both RNA and DNA can be used as the template, although RNA must first be transcribed to cDNA.
2. Prepare a reaction tube containing two oligonucleotide primers, one labelled with a fluorescent dye.
3. When hybridized to DNA, the primers are oriented to allow chain extension towards each other (one primer on the sense strand, and one on the anti-sense strand).
4. The tube also contains all the reagents necessary for chain extension, along with one chain-terminating ddNTP.

Protocol 18. *Continued*

5. Initiate the reaction with the addition of patient sample genomic DNA and a thermostable DNA polymerase.
6. The reaction mixture is thermally cycled, primers hybridize to template DNA, and oligonucleotides are generated which terminate at different positions across the target DNA sequence.
7. The vast majority of extension products end with a chain-terminating nucleotide, while a minority of the reaction products are extended far enough to serve as a template for hybridization of the other primer.
8. The reaction proceeds through 40 cycles, generating high levels of chain-terminated reaction product from each primer.
9. Detect the CLIP reaction products using an automated DNA sequencing system.
10. Compare the sequence to a user-defined database of known HIV sequences, and changes of particular interest.
11. Changes in a particular patient's viral sequences can easily be tracked.

In addition to identifying drug-resistant mutants, it is important to test new compounds for their antiviral effects against known drug-resistant mutants. Such mutants can be obtained from the AIDS reagent project (`http://www.nibsc.ac.uk/catalog/aids-reagent/` for Europe, or `http://www.aidsreagent.org/main.phtml` for the United States), but these mutants present their own difficulties in culture and testing. One such mutant has been isolated and propagated in HeLa cells expressing CD4. However, these strains will not grow in most cell lines used for cultivating HIV, including C8166 cells as used in the protocols above. In fact, propagation of the virus is restricted to MT2 and HeLa CD4-positive cells. A modification of the plaque reduction assay is used for their titration, and subsequently in antiviral sensitivity assays (30). A protocol is given below.

Protocol 19. Testing of antiviral compounds against AZT-resistant strains of HIV-1

Equipment and reagents

- HIV-1 mutant 105A, sensitive to AZT
- RPMI 1614 containing fetal calf serum, 2 mM L-glutamine, 100 units ml^{-1} penicillin, 100 $\mu g\ ml^{-1}$ streptomycin, and Geneticin (selects for CD4 antigen), and 100 $\mu g\ ml^{-1}$ hygromycin (selection for the BKLTRlac episome) (Sigma, Life Technologies, ICN Flow, etc.)
- HIV-1 105F, resistant to AZT (possesses mutations in RT at 67, 70, 215, and 219)
- MT2 cells
- Indicator cell line: HeLa CD4 +ve pBK LTR-*lac* carrying the *E.coli* β-galactosidase gene under the control of the HIV LTR
- X-gal (5-bromo-4-chloro-3-indoyl-β-D-galactopyranoside) (Sigma)

Method

1. Prepare confluent 96-well plates of the indicator cells.
2. Remove the supernatant, and add 100 μl of a predetermined dilution of the 105A supernatant and the 105F-infected MT2 cells.
3. Incubate the plates at 37 °C in 5% CO_2.
4. Examine the plates for syncytia after 3–5 days.
5. If required, remove the supernatant for confirmatory testing such as gp120 ELISA.
6. Stain the cells with X-gal, and then wash with PBS.
7. Add 100 μl of PBS containing 200 μg ml^{-1} of X-gal, 3 mM potassium ferrocyanide, 1.3 mM $MgCl_2$, pH 7.4.
8. Incubate the plates at 37 °C for 4 h or more to allow blue foci to develop.
9. Fix the cells with 10% formaldehyde in PBS for microscopic examination and counting.

7. Conclusions

Antiviral chemotherapy had a secure scientific beginning over four decades ago, with the discovery of the thiosemicarbazone group of molecules that were effective inhibitors of poxviruses. Marboran was subsequently tested in large clinical trials in India, and shown to prevent the onset of smallpox (31). Many fundamental discoveries were made in these early poxvirus investigations which were subsequently applied to later drugs, active against herpes, influenza, and HIV. Thus there were clear structure–activity relationships, in the sense that minor chemical modifications of the thiosemicarbazone resulted in dramatically decreased or increased antiviral effects. An important discovery was the very significant prolongation by the drug of the life of mice infected intracerebrally with poxviruses. Yet titration of viral infectivity of the brain demonstrated that only a 0.5 log reduction in virus infectivity was sufficient to ensure clinical recovery of the animal. Thus the clinical effects of the drug were not a result of complete inhibition of viral replication. Drug-resistant poxvirus mutants were isolated in cell culture and in animal model infections. Nevertheless, the drug clearly had a clinical effect in preventing smallpox infection in large trials in India. Modern antiviral chemotherapists still grapple with these same problems, but nowadays with viruses such as HIV or hepatitis B and C.

The extensive *in vitro* and *in vivo* studies with acyclovir over the last two decades have also illustrated that careful laboratory work in cell culture and animal models can be very predictive of clinical effectiveness. Acyclovir has become a gold standard to which other antiviral compounds are compared, not only scientifically but also economically. Early laboratory investigations of inhibition of HSV by acyclovir clearly established a mode of action, perhaps for the first time for an antiviral drug. This pioneering scientific work also highlighted the importance of viral enzymes as targets, in this case the herpes DNA polymerase enzyme. The newer nucleotide-sequencing techniques permitted analysis of drug-resistant herpes viruses to pinpoint mutations associated with resistance. Detailed comparison of acyclovir with a related drug, pencyclovir, also highlighted how small technical changes in cell culture and small changes in experimental design could alter the perspective of antiviral efficacy. Thus, although both drugs appear equally effective in the clinic, yet they have very different properties when they interact with virus-specific DNA polymerase and thymidine kinase enzymes.

It perhaps could have been anticipated that with the advent in the community of a completely new virus, HIV-1, the commercial scientists who had amassed ten years of laboratory experience with acyclovir would be the first to identify a new inhibitor, namely AZT. The scientific race to develop inhibitors of HIV has highlighted the potential of antiviral drugs. Also unprecedented has been a relaxation of government regulations enabling fast-track development of inhibitors of HIV-1. There are many examples now of new nucleoside and non-nucleoside substrate analogues of reverse transcriptase which proceeded straight from *in vitro* experiments in the laboratory, much as we have described above, into the clinic. The studies with HIV-1 have also highlighted the importance of precise diagnostic tests and PCR methodologies in kit form to quantify viral genomes in a patient. A key clinical test now in an HIV-infected patient is to show that a particular drug or combination of HIV drugs can ameliorate viral load, as quantified by the number of genome copies present in the plasma. A novelty, at least for antiviral chemotherapists, is that of drug combinations. Thus, HIV patients on highly active anti-retroviral therapy now receive three antivirals, AZT, another nucleoside analogue, and a protease inhibitor.

Six years ago, Von Itzstein (3) described a new inhibitor of the influenza neuraminidase (NA) enzyme, a viral target which had remained 'dormant' for years. Palese used the classical methods described in this chapter to evaluate a group of molecules related to this inhibitor (32). This original compound inhibited influenza plaque formation and caused budding viruses to aggregate at the cell surface, which would be expected if, as believed, the viral enzyme functioned to facilitate virus release. The novel twist was to modify the original molecule, which had failed to exert any effect in the ferret or mouse influenza model, using the × ray crystallography data on NA (33). The newly modified drug bound more tightly to the NA enzyme, inhibited replication of both influenza A and B viruses *in vitro* (3), in mice and in ferrets, as well as in volunteers and patients. Further modifications led to a molecule and pro-drug, which could be administered orally (5). Classic cell culture and animal model systems developed twenty to forty years ago were undoubtedly key elements in the rapid progress of these drugs towards the healthcare industry.

Virology undoubtedly has a molecular future, and viral enzyme targets in the form of rapid ELISA tests will be used to screen the millions of molecules already synthesized in the compound libraries of the world's leading pharmaceutical companies. But the classical virological cellular and animal model systems will still be the crucial turnstile through which selected drugs will pass *en route* to clinical use.

References

1. Hasegawa, T., Kurokawa, M., Yukawa, T., Horii, M., and Shiraki, K. (1995). *Antiviral Res.,* **27,** 271.
2. Craig, J., Duncan, I., Hockley, D., Grief, C., Roberts, N., and Mills, J. (1991). *Antiviral Res.,* **16,** 295.
3. von Itzstein, M., Wu, W., Kok, G., Pegg, M., Dyason, J., Jin, B., Van Phan, T., Smythe, M., White, H., Oliver, S., Colman, P., Varghese, J., Ryan, J., Woods, J., Bethell, J., Hotham, J., Cameron, J. and Penn, C. (1993). *Nature,* **363,** 418.
4. von Itzstein, M., Dyason, J., Oliver, S., White, H., Wu, W., Kok, G., and Pegg, M. (1996). *J. Med. Chem.,* **39,** 388.
5. Mendel, D., Tai, C., Escarpe, P., Li, W., Sidwell, R., Huffman, J., Sweet, C., Jakeman, K., Merson, J., Lacy, S., Lew, W., Williams, M., Zhang, L., Chen, M., Bischofberger, N., and Kim, C. (1998). *Antimicrob. Agents Chemother.,* **42,** 640.
6. Oxford, J., al-Jabri, A., Stein, C., and Levantis, P. (1996). In *Methods in enzymology* (eds. Kuo, L. C., Olsen, D. B., and Carrolls, S. S.). Vol. 275, p. 555. Academic Press, London
7. Lapatto, R., Blundell, T., Hemmings, A., Overington, J., Wilderspin, A., Wood, S., Merson, J., Whittle, P., Danley, D., Geoghegan, K., Hawrylik, S., Lee, S., Scheld, K., and Hobart, P. (1989). *Nature,* **342,** 299.
8. Tong, L., Pav, S., Pargellis, C., Do, F., Lamarre, D., and Anderson, P. (1993). *Proc. Natl. Acad. Sci. USA,* **90,** 8387.
9. Reichman, R., Badger, G., Mertz, G., Corey, L., Richman, D., Connor, J., Redfield, D., Savoia, M., Oxman, M., Bryson, Y., Tyrrol, D., Pomfrey, J., Creagh-Kirk, T., Keeney, R., Ashikaya, T., and Dolin, R. (1984). *J. Am. Med. Assoc.,* **251,** 2103.
10. Yarchoan, R., Klecker, R., Weinhold, K., Markham, P., Lyerly, H., Durack, D., Gelmann, E., Lehrman, S., Blum, R., and Barry, D. (1986). *Lancet,* 1, 575.
11. Nakashima, H., Matsui, T., Harada, S., Kobayashi, N., Matsuda, A., Ueda, T., and Yamamoto, N. (1986). *Antimicrob. Agents Chemother.,* **30,** 933.
12. Tobita, K., Sugiura, A., Enomote, C., and Furuyama, M. (1975). *Med. Microbiol. Immunol.,* **162,** 9.
13. Herrero-Uribe, L., Mann, G., Zuckerman, A., Hockley, D., and Oxford, J. (1983). *J. Gen. Virol.* **64,** 471.
14. Govorkova, E., Murti, G., Meignier, B., de Taisne, C., and Webster, R. (1996). *J. Virol.,* **70,** 5519.
15. Pauwels, R., Balzarini, J., Baba, M., Snoeck, R., Schols, D., Herdewijn, P., Desmyter, J., and De Clercq, E. (1988). *J. Virol. Meth.,* **20,** 309.

16. Weislow, O. S., Kiser, R., Fine, D., Bader, J., Shoemaker, R. H., and Boyd, M. (1989) *J. Natl. Cancer Inst.,* **81,** 77.
17. Oxford, J., Logan, I., and Potter, C. (1970). *Nature,* **226,** 82.
18. Smith, W., Andrewes, C., and Laidlaw, P. (1933). *Lancet* ii, 66.
19. Ruprecht, R., O'Brien, L., Rossoni, L., and Nusinoff-Lehrman, S. (1986). *Nature,* **323,** 467.
20. Hirsch, V., Olmsted, R., Murphey-Corb, M., Purcell, R., and Johnson, P. (1989). *Nature,* **339,** 389.
21. Agy, M., Frumkin, L., Corey, L., Coombs, R., Wolinsky, S., Koehler, J., Morton, W., and Katze, M. (1992). *Science,* **257,** 103.
22. Bobrowski, P., Capiola, R., and Centifanto, Y. (1991) *Int. J. Dermatol.,* **30,** 29.
23. Phillpotts, R., Welch, M., Ridgeway, P., Walkland, A., and Melling, J. (1988). *J. Biol. Stand.,* **16,** 109.
24. Landry, M., Myerson, D., and Bull, C. (1992). *Intervirology,* **34,** 169.
25. Myerson, D., and Hsiung, G. (1983). *Proc. Soc. Exp. Biol. Med.,* **174,** 147.
26. Stanberry, L., Kern, E., Richards, J., and Overall, J. C. J. (1985). *Intervirology,* **24,** 226.
27. Freeman, D., Sacks, S., De Clercq, E., and Spruance, S. (1985). *Antiviral Res.,* **5,** 169.
28. Darby, G., Field, H., and Salisbury, S. (1981). *Nature,* **289,** 81.
29. Hardy, W. (1992) *Am. J. Med.,* **92,** 30S.
30. Larder, B., Chesebro, B., and Richman, D. (1990). *Antimicrob. Agents Chemother.,* **34,** 436.
31. Bauer, D., Vincent, L., Kempe, C., and Downie, A. (1963). *Lancet,* **2**, 494.
32. Palese, P., and Schulman, J. (1977). In *Chemoprophylaxis and virus infections of the respiratory tract* (ed. J. Oxford), Vol. 1. p. 189, CRC Press, London.
33. Varghese, J., Laver, W., and Colman, P. (1983). *Nature,* **303,** 35.

9

Plant virus culture

E. P. RYBICKI and S. LENNOX

1. Introduction

Plant viruses are distinct from other major host-related groupings of viruses, such as animal or bacterial viruses, in that they are mostly single-stranded messenger-sense ss(+)RNA viruses, and mostly have simple, non-enveloped virions. There are a large number of types of simple isometric viruses, such as the *Bromoviridae*, *Tombusviridae*, and *Comoviridae*, as well as a wide variety of flexuous filamentous or rodlike types with helical capsid symmetry, such as the *Potyviridae*, *Closteroviridae*, and *Tobamoviruses* (1). There are some exceptions, notably with virus families that span different groups. Thus there are the enveloped ss(-)RNA plant *Rhabdoviridae* and *Bunyaviridae*, and the naked isometric dsRNA-containing plant *Reoviridae* (all of which may replicate in insect vectors and in insect cell cultures); there are two distinct groups of naked isometric ssDNA viruses with circular genomes (*Geminiviridae* and *Nanoviruses*) and two distinct groupings of plant pararetroviruses (*Caulimoviridae* and *Badnaviruses*) (2). While plant viruses as a group may not be as generally labile as their animal counterparts, given fewer enveloped viruses, it may be seen that there is considerable diversity, meaning that culture methods and methods for purification are similarly diverse.

Another major characteristic of plant viruses is their seeming lack of specific uptake mechanisms, in plants at least. Unlike most other viruses, plant viruses do not appear to have specific attachment proteins for the recognition of and binding to specific receptors on the surface of target cells. This has to do with the very different nature of the plant cell wall, compared with animal and bacterial cells: plant cells have thick, rigid, cell walls, composed largely of cellulose, whose dimensions are far greater than the sizes of viruses which infect them. Thus, unlike bacterial viruses, plant viruses cannot utilize 'injection' techniques to enter cells. Since plant cell membranes, unlike those of animal cells, are not exposed to the outside world, plant viruses do not appear to interact with them as part of the infection process. Instead, plant viruses enter cells from the outside either via transient breaks in cell wall and membrane (simple mechanical or vector-feeding damage), or via a specific injection process mediated by an insect vector.

The mode of transmission of viruses affects their concentration and localization in plants. For example, mechanically transmitted viruses (e.g. *Bromoviridae*, *Tobamoviruses*) tend to reach very high concentrations in most tissues of a plant (up to 4 g kg^{-1} of plant). This is necessary for survival, as it guarantees that a large number of virions will be present for onward transmission by whatever non-specific means presents itself. On the other hand, viruses which are introduced into plants via insect vectors with piercing mouthparts tend to be limited in their multiplication to phloem elements, which are preferred target tissues for insect feeding. Consequently these viruses (e.g. *Luteoviruses*, *Geminiviridae*) reach only very low concentrations (mg kg^{-1}) in whole plants. The route of transmission and the tissue tropism can profoundly affect the culture of plant viruses, in that very different purification procedures must be used for low-yielding as opposed to high-yielding viruses, and transmission of the viruses to new plants depends to a great extent on how well natural transmission mechanisms can be reproduced or mimicked.

A further effect of the structure of plant cells on plant virus replication cycles is the limitation on cell-to-cell and longer distance movement of viruses. Live plant cells interconnect only via specific discontinuities in the cellulose walls. The most numerous of these are plasmodesmata, which are complex structures filled with membrane-derived processes continuous with the endoplasmic reticulum. These are gated intercellular channels, which limit the passage of molecules between cells, and certainly do not admit particles as large as virions. Plant viruses have therefore evolved specific movement functions, mediated by one or more virus-specified proteins, which interact with the plasmosdesmatal machinery so as to increase the pore size, and allow specific transport of virus nucleoprotein complexes.

Plant viruses present a very different problem from that of animal or bacterial viruses when it comes to culture, as plant tissue-culture systems are far less routinely used. This is a function both of the problems inherent in setting up such systems, and the problems of infecting them with plant viruses. For example, while it is possible to grow suspension or solid callus cultures of many different types of plants, these cells all have thick cellulose walls, are difficult to infect with viruses by any simple procedure, and would limit virus infection to the cells inoculated. It is also possible (although complicated and expensive) to remove the cell walls of leaf-derived or culture-derived cells, and infect them by relatively simple procedures. These protoplasts may not be cultured on solid media as for plaque assays, and are very often not viable in terms of surviving more than a few days, let alone multiplying. Procedures for preparation and infection of plant protoplasts with plant viruses (3, 4) are covered elsewhere in this series. Accordingly, the rest of this chapter will focus largely on whole plant culture of plant viruses as opposed to cell culture methods; however, applicable techniques for the latter will also be discussed.

2. Isolation of plant viruses

Virus isolation includes both the biological isolation of a specific virus, and the physical purification processes used to extract it from infected tissues. Given the wide diversity of plant viruses, and differences in modes of transmission, it is obvious that no one technique can be prescribed for their isolation or purification. However, given similarities in their primary hosts, it is possible to lay down some general principles, and to establish some useful general protocols. For a more detailed account of specific procedures, there is still no better general source than Francki in Kado and Agrawal's *Principles and Techniques in Plant Virology* (5). Roger Hull (6) has also covered this ground very adequately in general terms; we shall therefore deal with specific practical points of general interest.

2.1 Virus isolation

Biological isolation of a plant virus is usually a procedure which is analogous to isolation of non-tissue culturable animal viruses such as hepatitis B virus: the effective lack of useful cell culture systems means that all procedures have to be carried out on whole plants, with all of the accompanying uncertainties inherent in any whole-organism experiment. For example, certain plants may contain very different amounts of polyphenols and other substances at different points in their growth cycles, or when cultivated under different conditions, which may adversely affect replication of any virus culture experiment. Similarly, different varieties of a given plant may have very different susceptibilities to the same virus (5). Another significant problem concerns obligately vector-transmitted viruses. In this case, non-viruliferous or non-infected cultures of the vector also usually have to be cultured in the laboratory before any virus work can occur. Recent developments in inoculation techniques have, however, lessened some of the need for insect and other cultures (*Protocol 1*).

The effects of plant viruses on their hosts are an important factor in their biological isolation. For instance, viruses which cause only systemic symptoms, such as generalized yellowing or mottling, would have to be isolated by end-point dilution experiments on groups of plants, with the hope that the single plant infected at the lowest effective dilution was the result of a single infective event. However, viruses which cause localized chlorotic or necrotic lesions on leaves of their hosts can be isolated by a procedure very similar to the plaque purification technique for bacterial and animal viruses: single well-separated lesions can be cut from a leaf, the virus extracted, and re-inoculated to another series of plants to repeat the process. It might be noted that the local-lesion assay predated animal virus cell culture by many years: it was first described in 1929 (7), whereas animal viruses were only successfully plaque-cultured in the 1950s (8).

The ideal situation for plant virus isolation would be to have a mechanically transmissible virus with an established local-lesion assay host, for ease of isolation, as well as a well characterized propagation host, from which large quantities of virions can be extracted. Viruses such as tobacco mosaic *Tobamovirus* (TMV) have been very well studied over many years because of just such properties. The worst case would be to have an obligately insect- or other vector-transmitted virus, with a narrow host range, and which reaches only very low concentrations in any host. Factors such as these are why viruses such as the obligately aphid-transmitted barley yellow dwarf *Luteoviruses* (BYDV), and whitefly-transmitted tomato yellow leaf curl *Begomovirus* (TYLCV, family *Geminiviridae*), have not been well studied until recently. Most real-life situations fall somewhere between these two extremes: for example, there are many viruses which are non-obligately vector transmitted, and which although they do not cause local lesions on any host, have suitable assay and propagation hosts, and can be purified in appreciable quantity. Viruses in the genus *Cucumovirus* (family *Bromoviridae*) and genus *Potyvirus* (family *Potyviridae*) would be excellent examples. They also represent two of the single largest groupings of viruses affecting plants, which is fortunate when it comes to culture!

Protocol 1. Biological isolation of a mechanically transmissible plant virus

This protocol assumes an unknown virus that is mechanically transmissible in a herbaceous plant. The appropriate range of plants that may be used for inoculation depends strongly on the virus(es) one is working with.

Equipment and reagents

A differential panel of plants used in our laboratory consists of:

- *Gomphrena globosa*
- *Datura stramonium*
- Squash (*Cucurbita pepo*) varieties (e.g. Long Green Bush)
- Various *Nicotiana* spp. (e.g. especially *benthamiana*; also *tabacum* cvs. Xanthi and Soulouk, *glutinosa, clevelandii*)
- *Chenopodium* spp. *hybridum, quinoa, amaranticolor*
- Cucumber (*Cucumis sativus*) varieties
- Garden bean (e.g. *Phaseolus vulgaris*) varieties
- *Petunia hybrida* (good general host)
- Zinnias (*Zinnia elegans*)
- Inoculation buffer: 0.05–1.0 M phosphate buffer, pH 7.0, with Celite®[a] (5 g per 100 ml; Celite Corporation, CA, USA)

A. *Inoculum preparation*

1. Homogenize fresh infected plant leaves or tissue with a mortar and pestle in a volume of inoculation buffer equivalent to twice the weight of the leaves, in the presence of some Celite®. If leaf tissue is tough, use some carborundum. When using lyophilized or air-dried material, use more buffer. Alternatively, make dilutions of purified virus in appropriate buffer, and add Celite®.

B. *Plant inoculation*

1. Gently rub upper surfaces of young leaves of plants at the 2–4 leaf stage with a cotton wool pad soaked in inoculum containing Celite®/ carborundum, ensuring that the leaf surface is wetted. Alternative inoculation methods include: inoculation with an artist's airbrush (9); inoculation with a particle gun (10); vascular puncture of seeds (11).
2. Leave plants for 10 min–1 h, then rinse with water from a watering can with a shower rose.
3. Place plants in dark, or away from light, for 1–2 days, then treat normally.
4. Monitor for development of symptoms, and score numbers of plants infected/those inoculated, and type and extent of symptoms.

C. *Virus isolation*

1. In the event of one of the experimentally determined hosts showing a local or even chlorotic lesion response, repeat the transmission at higher dilution (to isolate lesions from one another), and serially transmit to the same host (also to a host range) from homogenized single lesions. This is a good means of biological isolation of viruses from mixtures.
2. Where no lesion-producing hosts are found, inoculate several individual plants of the host giving the clearest systemic symptoms with serially diluted virus inoculum: single plants infected with the lowest infective dilution may contain single infections (test by onward transmission to a host range, note changes in range and symptoms).
3. Examine a range of infected plants by leaf-dip electron microscopy (negative staining of particles adsorbed onto coated grids from a droplet of sap or clarified leaf extract: see Chapter 5). Transmit onwards from those containing only one class of particles.

D. *Virus storage*

Viruses in infected leaf tissue may often be stored almost indefinitely, if dried appropriately. This may take the form of lyophilization, benchtop vacuum desiccation on silica gel, or simple air-drying of freshly-picked leaves in a commercial food dryer. In all cases, leaf tissue should be gently ground after drying, and immediately stored in sealed vials in between layers of dry paper towel, with self-indicating silica gel. These are best stored in a refrigerator (4°C), but can be stored at room temperature (for years) if virus particles are stable.

[a] Alternative buffer systems/additives should be used where plant homogenates contain large amounts of polyphenols or tannins, or other infection inhibitors: for example, 0.1 M diethyldithiocarbamate is a polyphenol oxidase inhibitor; 0.1 M ascorbate is an anti-oxidant (5).

2.2 Virus purification

There are probably as many (or more) plant virus purification techniques as there are plant viruses, depending as they do on differential stability of different viruses in different buffers, at different pHs, in extracts of different plants. A useful recent reference is (12), which contains a large number of modern plant virus purification techniques. Hull (6, and references therein) has previously summarized all the salient features of a typical plant virus purification protocol. These are, briefly:

- Extraction from the plant (homogenization of plant tissue, solubilization of virions: Section 2.2.1)
- Clarification of the extract (removal of cell debris, pigments, polyphenols: Section 2.2.2)
- Concentration/enrichment of the virus (ultracentrifugal pelleting/precipitation: Section 2.2.3)
- Final purification (differential ultracentrifugation, isopycnic or rate-zonal centrifugation, preparative electrophoresis: Section 2.2.4)

These classical methods work, with modification, with most plant viruses that have been characterized to date. Modern alternatives exist, however, which may allow the characterization of even the most recalcitrant or unstable plant viruses: these entail the isolation of replicative forms of the genomic RNA or DNA of a plant virus, and/or *in vitro* amplification of all or part of the genomes as DNA or cDNA (Section 2.3).

2.2.1 Extraction of virus

Extraction methods almost always involve homogenization of plant tissue in a suitable buffer system. The buffer should be chosen so as to maximize the stability of the virions (5), this being determined either biologically, using purified virions or fresh inocula from infected leaves, or physically by electron microscopy or analytical centrifugation (13). Suitable methods for particular viruses may be found in the *Descriptions of Plant Viruses* series (14). Homogenization can be done on a small scale, in a mortar and pestle, or by using kitchen or industrial appliances such as Waring-type blenders. Plant material is generally used fresh; although dried or frozen material can be used, yields are generally much reduced. The procedure should be done as quickly as possible, and material should be kept cold (4 °C). Homogenates are generally strained through cheesecloth, or other suitable coarse material.

2.2.2 Clarification of extracts

Simple clarification consists of low-speed centrifugation (10 000 *g* for 10 min), again usually in the cold (4 °C). This is aided by use of low pH buffers, which denature many plant proteins. However, these may not be suitable for many viruses. Organic solvents such as chloroform or *n*-butanol have often been used to remove pigments, plant proteins, etc. However, worker safety considerations should preclude the large-scale use of solvents. Filtration through Celite®-charcoal pads (made on a filter using a 1:1 w/w slurry, in appropriate buffer, of Celite and activated charcoal: Adcoa Corp., CA, USA) in Buchner funnels is an extremely simple means of decolorization of plant extracts (5).

2.2.3 Concentration of extracts

Ultracentrifugation (50 000–100 000 g for 2–3 h) is the simplest means of concentration of virions. Pelleted components are then resuspended in appropriate buffer, and further enriched by cycles of differential ultracentrifugation. This works especially well for viruses which yield well, and which have stable particles which withstand repeated pelleting. Another very useful means of concentration of an unknown virus, which may be low-yielding, meaning large volumes of extract have to be concentrated, is the use of polyethylene glycol (PEG) MW 6000–10 000. This is added to the extract to produce concentrations of 2.5–10% (w/v), with or without NaCl (0.1–1.0 M). The extract can be left overnight at 4 °C, the precipitate is then collected by centrifugation, and is resuspended in a small volume of buffer.

2.2.4 Purification

Stable viruses may be purified to a high degree by simple differential ultracentrifugation. Alternatively, banding of virions by isopycnic gradient centrifugation (usually CsCl, alternatively Cs_2SO_4) may be used to obtain extremely pure preparations. More labile virions can be centrifugally purified by banding in rate-zonal gradients (usually sucrose, 10–40%), or sedimentation through a sucrose cushion. An older, yet still viable, technique is rate-zonal electrophoresis in a sucrose gradient: this may still be very useful in obtaining purified fractions of viruses which cannot be separated by centrifugation techniques, but which do differ in surface charge (15, 16).

2.3 Purification of virus genomic nucleic acid

While purification of virions may require a different method for every virus, it is possible to purify virus-derived nucleic acids from plants or from cells by relatively simple, standard techniques. It is also possible to isolate genomic nucleic acid (NA) from purified virions. However, inasmuch as the virion purification process inevitably leads to significant losses of particles, it may be more rewarding (and considerably quicker) to isolate virus NAs directly from plant material. While most of the reason to isolate virus NAs is for making hybridization probes for the viruses, it is also possible to isolate whole infectious genomes for most plant viruses.

A great advantage of the replication cycle of ss(+)RNA viruses (which includes most plant viruses) is that one can isolate, by very simple methods, significant quantities of dsRNA corresponding to whole genome replicative forms or intermediates from infected plants. While at first it was thought this only worked well for some viruses (17), such as cucumber mosaic virus (CuMV), it later became apparent that genome-sized dsRNAs could be isolated even from large single-component genome viruses such as potyviruses or closteroviruses (18).

It is also possible, and in fact desirable, in terms of obtaining clonable DNA, to isolate plasmid-like dsDNA replicative forms of the genomes of DNA plant viruses as diverse as caulimoviruses (19) (which have gapped circular dsDNA genomes), and geminiviruses and nanoviruses (20) (which have circular ssDNA genomes). These are often infectious, and may in fact be introduced back into plants by techniques such as biolistic inoculation, with or without prior cloning (21).

Polymerase chain reaction (PCR) amplification allows rapid production of large amounts of dsDNA corresponding to all or part of a RNA or DNA plant-virus genome. Modern techniques using more stable polymerases increasingly allow the production of full-length genomes even from viruses with relatively large genomes, such as potyviruses.

2.3.1 Virion nucleic acid purification

This has been well and recently covered (12), and little can be added for any practical purpose. Given, however, that one overriding reason to purify virus nucleic acids is to clone them as DNA or cDNA, an extremely useful technique described by Wyatt *et al.* (22) may help in obtaining cDNA clones from RNA viruses. The authors effectively used a reverse transcriptase (RT)-mediated stripping of coat protein from freeze-thawed virions to obtain full- or near-full-length cDNA clones from potyviruses, and from a wide variety of other spherical and filamentous viruses. They mention that it can be incorporated into an immunocapture-RT-PCR procedure, meaning that there may be no need to purify virions at all (23).

2.3.2 dsRNA purification

While techniques for virus dsRNA purification from plant tissue are very well established (18, 24), the fact that plant-derived virus dsRNA may in many cases be used as a long-term means of storing an infective virus isolate merits a detailed protocol. This protocol is derived from Gehringer (25) and Hall (26), and is adapted from that of Valverde (18). It utilizes the differential adsorption of different nucleic acids to cellulose in the presence of ethanol as a means of fractionation of dsRNA from total plant nucleic acids.

Double-stranded RNA isolated with the simple technique described above may be stored indefinitely, either under 70% ethanol on the bench, or in TE at 4°C, or frozen in TE. The infectivity of dsRNA from CuMV was tested in this laboratory; all plants inoculated with denatured dsRNA became infected, while plants inoculated with native dsRNA showed no symptoms at all (25). This finding may be extremely useful to other laboratories, as it represents a novel means of indefinite storage of virus inocula. It is by no means certain that it works with all viruses; however, the fact that it works with something that is multicomponent like CuMV means there is probably very broad applicability of the technique.

Protocol 2. dsRNA extraction from infected plant tissue and re-inoculation into plants

Equipment and reagents[a]

- Autoclaved 50 ml centrifuge tubes (e.g. Sorvall SS-34 polypropylene tubes) with screw caps
- 50 ml disposable sterile plastic syringe barrels
- Small autoclavable ceramic mortars and pestles, or electric coffee grinder (e.g. Bosch MKM 8000)
- Retort stand and clamp, plastic tubing (for syringe outlet), and tube clamp
- Submarine horizontal 'mini' gel electrophoresis unit and power pack
- Liquid nitrogen and insulated container
- Thick gloves (e.g. heavy-duty oven gloves)
- Whatman CF-11 cellulose powder (or equivalent), 2.5 g per extraction
- Saline-Tris-EDTA buffer (STE), pH 6.8: 0.1 M NaCl, 0.05 M Tris, 1 mM EDTA; 20 ml per extraction
- STE containing 16% (v/v) reagent-grade ethanol (STE-EtOH), 100 ml per extraction
- 3M sodium acetate (NaAc), 0.5 ml per extraction
- Tris-Borate-EDTA (TBE) buffer, pH 8.0: 10x stock is 0.45 M Tris, 0.45 M boric acid, 0.01 M EDTA
- STE-saturated phenol, 10 ml per extraction
- Tris-EDTA (TE) pH 7.5: 10 mM Tris, 1 mM EDTA; 0.5 ml per extraction
- Reagent-grade 95% ethanol
- Dimethyl sulfoxide (DMSO)
- 10% (w/v) sodium dodecyl sulfate (SDS), 1 ml per extraction
- Ethidium bromide (EthBr), 10 mg ml^{-1} in water: store in the dark
- 2% (w/v) bentonite in water (resuspend before use); 0.5 ml per extraction
- Young tobacco plants infected with an RNA virus (e.g. CuMV, TMV)
- Young (e.g. four-leaf stage) healthy tobacco plants
- Celite® (Celite Corporation, CA, USA)
- Agarose

Optional (see A.9 below):

- TNM buffer: 0.1 M Tris, 1 M NaCl, 0.1 M $MgCl_2$, 50% (v/v) glycerol, pH 7.8
- DNAse, 1 mg ml^{-1} in TNM buffer
- Proteinase K, 5 mg ml^{-1} in TNM buffer
- $CaCl_2$, 0.45 M
- $MgCl_2$, 0.5 M
- LiCl, 8 M
- Chloroform–isoamyl alcohol ($CHCl_3$: isoamyl), 24:1 (v/v)

A. *dsRNA extraction*

1. Cut up leaf material with scissors or blade, then grind 5 g leaf tissue to a fine powder in the presence of liquid nitrogen, using either a small sterile mortar and pestle (note: use gloves!) or a coffee grinder cleaned with 70% ethanol. Note that the material should be kept powdered by the continued addition of liquid nitrogen until grinding is finished.
2. Add powdered leaf material to 8 ml of STE in a centrifuge tube, mix, and add 1 ml of 10% SDS, 0.5 ml of 2% bentonite, and 9 ml of STE-saturated phenol. Cap and shake tubes thoroughly at 4°C for 30 min.
3. Centrifuge the tubes (8000 *g*, 15 min), then transfer supernatant fluid to a new tube. Add ethanol to 16% (v/v) final concentration.

Protocol 2. *Continued*

4. Equilibrate 2.5 g CF-11 cellulose with 30 ml STE-EtOH in a 100 ml beaker at room temperature. Secure a 50 ml syringe barrel plugged with sterile glass wool vertically in a retort stand, with a short length of plastic tubing attached to the outlet, with a clamp to arrest liquid flow. Resuspend the cellulose and pour into the syringe barrel. Allow to drain.
5. Allow the extract to reach room temperature, then pour onto the cellulose column. Drain the column, then wash with 40 ml STE-EtOH. This step allows preferential adsorption of dsRNA to the cellulose.
6. Elute the dsRNA fraction with 10 ml STE, collected in a 50 ml centrifuge tube. The dsRNA may be precipitated at this stage by the addition of 20 ml ethanol and 0.5 ml 3 M NaAc, followed by centrifugation (8000 *g*, 30 min); incubation may be for as short a period as 30 min at room temperature, or overnight at 4°C, or a few hours at −20°C. Alternatively, the dsRNA solution can be made up to 16% (v/v) with ethanol, passed over a fresh cellulose column, eluted in 6 ml of STE, and precipitated as described above. Only high-dsRNA-yielding samples should be fractionated twice, however.
7. Precipitated dsRNA (note: pellet is usually invisible) may be washed with 70% (v/v) ethanol by simply covering the pellet and turning the tube around by 180° and re-centrifuging. Drain the pellet by tilting the tube away from the pellet location, and pouring off the liquid, then standing the tube upside down on a pad of paper towel for 10–20 min.
8. Resuspend the pelleted dsRNA in 0.1–0.5 ml TE. This may be stored indefinitely at 4°C.

Optional steps (not usually necessary, even for cDNA synthesis):

9. Mix 30 μl DNAse stock with 4 μl 0.45 M $CaCl_2$ and 30 μl proteinase K stock, incubate for 1 h at 37°C: this removes RNAse activity from the DNAse. Add 10 μl of this and 5 μl of 0.5 M $MgCl_2$ to 0.5 ml of dsRNA in TE. Incubate for 1 h at 37°C: this removes all DNA. Extract with 2 volumes of STE-phenol and 1 volume of $CHCl_3$:isoamyl alcohol by shaking, separate phases by centrifugation, and precipitate dsRNA with 2 volumes ethanol. Resuspend as before (A8).
10. Add one-third volume of 8 M LiCl to dsRNA solution, and store at 4°C for 16 h. Precipitate ssRNA by centrifugation (e.g. Eppendorf-type bench centrifuge, 10 000 r.p.m. for 20 min). Remove supernatant to new tube, and add one-third volume 8 M LiCl to supernatant (becomes 4 M LiCl), and repeat 4°C storage: this precipitates dsRNA, which is pelleted and resuspended as before (A8).

B. *Quantitation and purity determination*

The dsRNA may be quantified by spectrophotometry at 260 nm, using a specific absorption coefficient of 20 OD units μg^{-1}. Purity of the dsRNA may be conveniently estimated by means of agarose gel electrophoresis and staining with ethidium bromide.

1. Prepare 0.8–1.5% (w/v) agarose in 1 × TBE, pour into gel trays, add comb.
2. Mix dsRNA samples 1:2 (v/v) with loading buffer (TBE containing 10% glycerol and some bromophenol blue).
3. Load into gel slots (10–40 µl), subject to electrophoresis at (for example) 30 mA per 100 × 10 mm cross-section gel for 1–3 h, or until dye reaches near the end of the gel.
4. Place gel in TBE containing 50 ng ml^{-1} EthBr, shake gently for 30 min.
5. View gel on UV trans-illuminator (254 nm): RNA bands should be bright, but photographs may need exposures of 1–10 s to detect RNA in lower-yielding preparations.
6. To check whether bands are RNA or DNA, one can include RNAse A (10 mg ml^{-1} stock, 10 µl ml^{-1}) in the loading buffer, or soak gels in RNAse/DNAse-containing buffer for 60 min, and re-photograph.

C. *Inoculation of plants with denatured dsRNA*

1. Add DMSO to the dsRNA solution to a final concentration of 50% (v/v).
2. Heat in a closed vial in a waterbath at 80°C for 10 min, then transfer immediately to an ice bath.
3. Add 1 volume of 2% bentonite to samples, still on ice, mix well.
4. Dust leaves of young tobacco plants with Celite®, then rub with a sterile cotton wool pad soaked in dsRNA inoculum.
5. Rinse leaves with water, leave plants for development of symptoms (7–14 days).

[a] all buffers should be autoclaved where possible

2.3.3 Extraction of DNA genome viruses as plasmids

We have previously described this technique specifically for geminiviruses (Palmer *et al.* (20)). However, the technique is sufficiently broad-based in concept to be used for the circular ssDNA geminiviruses, the structurally similar nanoviruses, and the gapped circular dsDNA caulimoviruses and badnaviruses. The plasmid-like DNAs are covalently-closed circular (ccc) DNA molecules, which in their native form are complexed with host nucleosomes in minichromosome-like structures (26). Essentially, one performs a total nucleic acid (TNA) extraction from infected plants by whatever method yields the highest level of 'clean' TNA, and follows that with a plasmid-specific purification. A number of commercially available kits may be used: for example, QIAPrep (Qiagen GmbH, Hilden, Germany) or Nucleobond (Macherey-Nagel GmbH, Germany) or High Pure (Boehringer Mannheim) plasmid kits have all been used in our laboratory for geminivirus purification. The protocol given was developed by Fiona Hughes (27) in our laboratory, for maize streak virus (MSV) dsDNA purification. Plasmid-type DNA purified by this method is immediately suitable for cloning into plasmid vectors, or, if yield is sufficient, for restriction endonuclease digestion and RFLP analysis. It is also infectious, if inoculated appropriately (21,28,29).

Protocol 3. cccDNA purification from plants

Equipment and reagents

- Extraction buffer (0.1 M Tris-HCl, pH 7.0; 0.1 M NaCl; 0.1 M EDTA; 1% [w/v] SDS)
- Buffer-saturated phenol (Sambrook *et al.* (30))
- LiCl solution, 4 M
- Tris-EDTA buffer (TE, *Protocol 2*)
- Solution I (25 mM Tris-HCl, pH 8.0; 10 mM EDTA; 50 mM glucose)
- Solution II (0.2 M NaOH, 1% [w/v] SDS)
- Ethidium bromide solution (10 mg ml^{-1} stock)
- Solution III (5 M potassium acetate, pH 4.8)
- 10 mg ml^{-1} DNase-free RNase A (boil solution briefly to ensure that it is DNase-free)
- 50 ml centrifuge tubes (e.g. Sorvall SS-34 polypropylene tubes) with screw caps
- Mortar and pestle or coffee grinder (*Protocol 2*)
- Sephadex G-100 resuspended in TE buffer (*Protocol 3*)
- 10 ml column (e.g. 10 ml syringe barrel)

A. *Total nucleic acid (TNA) extraction*

1. Cut plant material into small pieces, freeze in liquid nitrogen for 5 min, and grind into a fine powder.
2. Add an equal volume of extraction buffer, and stir the mixture for 5 min at room temperature. Filter the homogenate through a single layer of cheesecloth.
3. Extract the filtrate once with half its volume of a 1:1 mix (v/v) of phenol:chloroform.
4. Centrifuge this mixture for 10 min (10 000 *g*), and remove the aqueous phase to a fresh centrifuge tube.
5. Mix the aqueous phase with an equal volume of propan-2-ol (isopropanol), and centrifuge immediately (15 min at 10 000 *g*).
6. Resuspend the crude TNA pellet in 4 ml TE buffer and reprecipitate, using 0.1 volume 4 M LiCl and 2 volumes 96% ethanol. The pellet should be resuspended in 1 ml TE.

B. *DNA purification*

This is a column purification which serves to rid the samples of residual protein and plant pigments not removed by phenol:chloroform extraction.

1. Pack a 10 ml column with Sephadex G-100, equilibrate with TE buffer.
2. Add 1 ml (or 1/10 column volume) of TNA in TE, elute with 400 μl aliquots of TE, collecting fractions in Eppendorf-type vials.
3. Assay spectrophotometrically at 260 nm: run a 230–310 nm scan for high-absorbing samples. Pool all samples with typical DNA/RNA absorption profiles.
4. Precipitate nucleic acids from pooled samples, using 0.1 volume 4 M LiCl and 2 volume 96% ethanol. Resuspend the pellet in 1 ml TE.
5. Remove RNA by addition of RNase A to a final concentration of 1 mg ml^{-1}. Re-extract with phenol:chloroform and re-precipitate DNA, or simply re-precipitate DNA.

C. *cccDNA purification*

A simple plasmid-type 'maxiprep' procedure is described: proprietary column techniques may also be used, as per manufacturers' instructions.

1. Resuspend the DNA in 2 ml of solution I in 50 ml centrifuge tube.
2. Add 4 ml of solution II, shake to mix.
3. Place tube on ice for 5 min. Add 1.5 ml of cold solution III, shake tube briefly, and keep for 5 min on ice.
4. Centrifuge at 10 000 *g* (e.g. Sorvall SS34 rotor) for 10 min at 4°C. Nucleic acid in the supernatant is precipitated with an equal volume of isopropanol, and pelleted by centrifugation at 10 000 *g* for 15 min. Resuspend the pellet in 4 ml TE.
5. Add CsCl (1.2 g per ml of solution) and EthBr (200 μl), and adjust the solution's refractive index to 1.396.
6. Centrifuge tube (10 000 *g*), and decant the supernatant into 5 ml Quick-Seal tubes, and seal. Centrifuge overnight at 55 000 r.p.m. in a VTi 65 rotor in a Beckman L8-70 ultracentrifuge at 15°C.
7. The plasmid-type band is recovered using a syringe and needle from the side of the tube, as described in Sambrook *et al.* (30). EthBr can be removed by exhaustive extraction with TE/NaCl-saturated isopropanol.
8. The DNA solution should be precipitated by addition of 2× volume of TE and 1× resultant volume of isopropanol.
9. The DNA pellet can be washed with 70% ethanol and resuspended in TE.

2.3.4 PCR amplification of plant virus genomes

While this is more generally useful as a diagnostic technique, when partial genome amplification is the norm, amplification and/or cloning of a whole virus genome is also of great use in virus culture, as it allows indefinite archiving of a plant virus as a DNA copy, and its propagation in bacteria as a clone. It is now almost a matter of routine for researchers to obtain infectious cDNA clones of RNA viruses, even those with divided genomes. It is easier, however, to obtain infectious clones or even direct PCR products for viruses with undivided genomes: the ssDNA geminiviruses are one case in point (25, 31). However, it is also possible for sobemoviruses (rice yellow mottle, RYMV (32)), and potyviruses such as potato virus Y (PVY (33)). In the case of circular DNA genomes (geminiviruses, nanoviruses, caulimoviruses, badnaviruses), back-to-back primers should be used: these can be abutting, or partly overlapping over a convenient restriction site.

For linear RNA genomes, primers representing the 5′ and 3′ terminal sequences must be used. Newer heat-stable polymerases and new protocols allow the amplification of cDNAs over 10 kb pairs (e.g. Boehringer Mannheim), meaning that whole genomes of potyviruses, etc. may be amplified in a single reaction. If the products of PCR amplification are to be cloned, then this should be in such a manner as to allow transcription from a specific promoter (e.g. CaMV 35S) at the native 5′ genome terminus, with termination at or as close as possible to the 3′ terminal base. Alternatively, the construct could be cleaved at the 3′ site before either *in vitro* transcription of RNA, or direct inoculation of appropriate DNA onto plant leaves (34). It is possible to construct infectious cDNA genomes of a potyvirus totally *in vitro*. Fakhfakh *et al.* (33) used dsDNA ‘megaprimers’ to amplify potato virus Y genomic RNA into a final PCR product containing a cauliflower mosaic virus 35S RNA promoter and a nopaline synthase terminator. Biolistic bombardment with a helium particle gun was used to inoculate the amplified product to detached tobacco leaves: up to 90% of bombarded leaves contained wild-type virus as assayed by immunoelectron microscopy and ELISA.

3. Tissue culture and plant viruses

Plant tissue culture is the backbone of plant biotechnology (35). Advances in agricultural biotechnology have generated much-needed new genetic variability, which has enabled improvement or early release of new cultivars of a number of crops. Potrykus and Spangenberg (36) have edited a laboratory manual focusing on the current major techniques for efficient, routine production of transgenic ‘model plants’, such as *Arabidopsis thaliana,* tobacco, and maize. Protocols given for techniques of interest here include *Agrobacterium*-mediated gene transfer; protoplast-based direct gene transfer, and biolistic DNA transfer to tissues. Direct gene-transfer methods include electroporation and polyethylene glycol (PEG)-mediated gene transfer. PEG-mediated transfection of protoplasts with a geminivirus, and biolistic inoculation of *Triticum monococcum* suspension culture cells, are also covered by Mullineaux (3).

However, older methods are also possible. In 1974, Hansen (37) reported the regeneration of a tobacco mosaic virus-infected tobacco callus culture, and as long ago as 1971, Moskovets *et al.* (38, 39) reported the infection of different suspension and callus cultured-cell systems with potato virus x and its genomic RNA.

Plant tissue culture involves the clonal multiplication (micropropagation) of plants *in vitro* by means of vegetative propagation. This includes axillary shoot production from tissues containing pre-existing meristems, adventitious shoot production following induction of adventitious meristems, and somatic embryogenesis. This may also include multiplication and regeneration from protoplasts and single cells, culminating in adventitious shoots or somatic embryos (40). George and Sherrington (41) and Kyte (42) have described thoroughly the fundamentals of plant culture media and methods. George *et al.* (43) compiled media formulations and uses, which can be adapted for a large variety of plant species.

We have used our experience with maize culture to describe some of the media and methods involved in initiating, selecting, maintaining, and regenerating maize callus and suspension cell cultures, and somatic embryogenesis, as all of these different tissue culture systems may be used for virus culture. Freeling and Walbot (44) have edited protocols describing maize culture and transformation in more detail. We also describe a biolistically mediated virus-transient-expression assay in non-regenerable maize Black Mexican Sweetcorn (BMS) suspension-cultured cells (Section 3.4).

BMS suspension-cultured cells are an extremely useful model system for transformation and gene expression studies (45), as the cells can readily be genetically transformed. BMS callus and suspension cultures are soft, friable, and homogenous in nature. They grow rapidly and are easily managed (46). Other protocols for using and preparing protoplasts from *Triticum monococcum* suspension cells, and making barley protoplasts, are described in this series by Mullineaux (3), and by Kroner and Ahlquist (4).

3.1 Callus initiation

Regenerable maize cultures are generally classified as type I or type II cultures, depending on (a) friability (how closely cells are physically associated to each other), (b) mode of regeneration, and (c) degree of differentiation (47). Type I cultures are non-friable, regenerate through embryogenesis or organogenesis, and are highly differentiated. Type II cultures are friable, regenerate through somatic embryogenesis, and are relatively undifferentiated, making it easier to establish suspension cultures, and to select for embryogenic callus *in vitro* (47). BMS cultures are available from a number of laboratories. For long-term maintenance, it is easiest to keep the cells on solid media (e.g. Murashige and Skoog 2SMS2D, ref. 48).

Protocol 4. Maintenance of BMS cells on solid media

Equipment and reagents

- 2SMS2D medium: 4.4 g l^{-1} MS salts (Sigma or Gibco); 0.25 mg l^{-1} each of thiamine-HCl, pyridoxine, and calcium pantothenate; 1.3 mg l^{-1} niacin; 0.13 g l^{-1} L-asparagine; 0.2 g l^{-1} myo-inositol; 2 mg l^{-1} 2.4-D; and 20 g l^{-1} sucrose. Adjust the pH to 5.8 with KOH.
- Phytagel (Sigma), or GELRITE (Sigma), or purified agar (Sigma)
- Suspension culture of BMS cells: American Type Culture Collection (ATCC 54022)

Method

1. Prepare 2SMS2D medium.
2. Add suspension-cultured cells.
3. Solidify the medium with 2 g l^{-1} Phytagel, 0.2 % GELRITE or with 7g l^{-1} agar.
4. Maintain the cultures free of microbial contamination by working with them in a sterile flow hood, and by using sterilized tools and media (45).
5. Keep cultures at 25 °C in the dark with bi-monthly transfers.

Maize type II culture establishment is genotype-dependent. Armstrong *et al.* (49) selected Hi-II germ plasm from the experimental genotype A188 × B73. This Hi-II germ plasm forms frequent and vigorous type II cultures, is publicly available, and is not as variable and dependent on environmental conditions as A188 and the F1 hybrid. Sellmer *et al.* (50) described culturing type II callus from inbred A188 and A188 derivatives. Dunder *et al.* (51) described culturing and transforming elite maize lines after microprojectile bombardment of immature embryos. Armstrong (47) initiated Hi-II regenerable callus from immature embryos that are approximately 1–2 mm in length, normally 8–12 days post-pollination.

Protocol 5. Regenerable callus initiation from Hi-II maize

Equipment and reagents

- Surface sterilising agent (3.5 % NaOCl)
- Callus initiation medium (2SN61D). This contains 4 g l^{-1} N6 salts (Sigma)(15); 2.88 g l^{-1} L-proline; 2 mg l^{-1} glycine; 1 mg l^{-1} thiamine-HCl; 0.5 mg l^{-1} pyridoxine-HCl; 0.5 mg l^{-1} nicotinic acid; 100 mg l^{-1} casamino acid; 20 g l^{-1} sucrose; and 1 mg l^{-1} 2.4-D (52).
- Phytagel (Sigma)
- Hi-II maize with immature cobs[a]

Method

Caution: work aseptically!

1. Surface-sterilize maize cobs in 3.5 % NaOCl for 30 min, followed by three washes of 10 min each in sterile, distilled water.

2. Excise embryos from each kernel, and place with the flat (embryonic axis) side down on 2SN61D.
3. Set pH to 5.8, and solidify with 2 g l^{-1} Phytagel.
4. Add 10 μM silver nitrate to the autoclaved medium. Incubate in the dark at 28 °C, and transfer to fresh medium fortnightly (10).

[a] Optimum embryo length is 1.0–2.0 mm. This should normally occur within 8–12 days of pollination, depending on health and growing conditions of donor plants (10)

Type II callus must be continually selected at each subculturing to maintain highly embryogenic, soft, and friable callus. It may show heterogeneity in embryoid structure, stage of embryoid development, or in colour, from translucent to light yellow. Hard callus tissue not showing embryoid structures, mucilaginous callus, or tissue showing highly advanced embryoid development, should be discarded.

3.2 Establishing a suspension culture

BMS suspension cultures can be initiated as shown in the protocol below.

Protocol 6. Establishment of BMS suspension culture from callus

Equipment and reagents

- 2SMS2D liquid medium (*Protocol 5*)
- 250 ml Erlenmeyer flasks
- Solid (callus) culture of BMS on solid medium (*Protocol 5*)

Method

Caution: work aseptically!

1. Add a 3 ml cell volume to 50 ml 2SMS2D liquid medium in flasks.
2. Incubate flasks on a shaking platform at 140 r.p.m., in the dark at 25 °C.
3. Maintain the cells by tenfold dilutions every week.

BMS can be genetically transformed by means of microprojectile bombardment of cells (46,53), or by direct gene transfer methods such as electroporation or polyethylene glycol (PEG)-mediated transformation of protoplasts (45). BMS is a useful model system, but the cells do not regenerate.

Establish a Hi-II cell suspension culture firstly by subculturing the callus on fresh solid medium weekly. This reduces embryoid development, and aids in dispersion of the callus in liquid medium (50).

Protocol 7. Establishment of Hi-II suspension culture from callus

Equipment and reagents[a]

- Callus initiation medium (2SN61D) (*Protocol 5*)
- Hi-II embryogenic callus culture
- 250 ml Erlenmeyer flasks

Method

Caution: work aseptically!

1. Select and dispense 1–2 g fresh weight of friable, soft, embryogenic callus into 40 ml liquid Hi-II callus-initiation medium (2SN61D) in 250 ml flasks.
2. Incubate these flasks on a shaking platform at 140 r.p.m. in the dark at 25 °C.
3. After one week, begin a 7-day wash cycle every third and seventh day for two to three weeks. Wash cells by removing the old medium, replacing it with fresh medium, and taking care to minimize removal of cells.
4. Once the Hi-II cells have grown to the extent that subculturing is required, begin a standard 7-day culture/selection cycle. Aim to produce small aggregates of highly cytoplasmic and actively dividing cells, rather than large, irregularly shaped, highly vacuolated cells.
5. Wash the cells once with fresh medium, then selectively pipette a 4–5 ml packed cell volume (PCV) of small cell aggregates into a new 250 ml flask containing 40 ml of medium. Adjust the PCV used for subculturing to fit the speed with which the cells grow.
6. Refresh the cells by removing the old medium, and replacing it with fresh medium on the third or fourth day of the culture/selection cycle.
7. Eventually omit the washing step, once the cultures remain actively dividing and produce a small amount of debris (50).

Protoplasts can be prepared (54) and cell lines can be cryopreserved (55) from suspension-cell cultures. Many genetic manipulations can be done using protoplasts. For example, genetic modification by incorporation of foreign DNA (direct gene transfer by PEG or electroporation), fusion with maize or other species, and generation of somaclonal variants (54). Cryopreservation of established cell lines reduces the need and expense of continually developing new lines, as cultures often lose the ability to regenerate large numbers of plants after one year (50).

3.3 Production of somatic embryos from regenerable type II cultures

Somatic embryos are very useful for the regeneration of transgenic maize (and other plants), but their use in virus culture is not well established. However, we include mention of the techniques because of the potential relevance of geminivirus-based expression systems in transgenic maize (56). The protocol below is based on that of Armstrong (47), who regenerates Hi-II and other type II cultures in three steps.

Protocol 8. Production of somatic embryos

Equipment and reagents[a]

- Phytagel (Sigma)
- Type II Hi-II (or other) highly embryogenic, soft and friable callus (*Protocol 5*).
- Regeneration medium 1 (2SN62D): this contains 4.4 g l^{-1} MS salts (Sigma); 1.3 mg l^{-1} nicotinic acid; 0.25 mg l^{-1} pyridoxine-HCl; 0.25 mg l^{-1} thiamine-HCl; 0.25 mg l^{-1} calcium pantothenate; 100 mg l^{-1} myo-inositol; 1 mM asparagine; 0.1 mg l^{-1} 2.4-D; 0.1 μM abscisic acid (ABA); and 20 g l^{-1} sucrose. Set pH at 5.8, and solidify with 2 g l^{-1} Phytagel.
- Regeneration medium 2 (6SN60D): this contains 4 g l^{-1} N6 salts (Sigma); 0.5 mg l^{-1} nicotinic acid; 0.5 mg l^{-1} pyridoxine-HCl; 1 mg l^{-1} thiamine-HCl; 2 mg l^{-1} glycine; and 60 g l^{-1} sucrose. Set pH at 5.8, and solidify with 2 g l^{-1} Phytagel.
- Regeneration medium 3 (2SN60D): Regeneration 1 (2SN62D) medium without 2.4-D.
- Peat:sand (1:1 w/v) sterilized growing medium.

Method

Caution: work aseptically!

1. Select embryogenic callus from a Type II culture.
2. Differentiate selected somatic embryos on Regeneration medium 1 (2SN62D). Leave embryos on medium for two weeks in the dark at 28 °C.
3. Enlarge and mature somatic embryos on Regeneration medium 2 for two weeks in the dark at 28 °C.
4. Germinate somatic embryos on Regeneration medium 3 for two weeks under a 16:8-hour light:dark photoperiod at about 25 °C.
5. Shoots formed on Regeneration 3 are rooted on the same medium, then planted out into peat: sand for hardening off under gradually reduced humidity, gradually increased light intensity, and increased soil richness.

[a] all reagents should be sterile (except soil).

3.4 Transient virus replication in callus or suspension cells

Techniques for plant virus assay or multiplication of viruses in plant cell culture systems are not particularly well worked out, except for protoplast systems (Section 1). The problems inherent in getting viruses or virus nucleic acids through the thick cell wall of cultured cells are bad enough; the fact that viruses cannot spread out of the cells once inside, again because of the cell walls, means that any assay of multiplication can only be of cells infected in the initial process. This rules out any plant cell equivalent of plaque assays, except for viruses like the Phycodnaviridae, which can both enter and leave unicellular and other algal cells by specialized mechanisms (57).

Protoplast systems for DNA and RNA virus multiplication assays are well discussed in a previous volume of this series (3, 4), and will not be further discussed here. We will describe a transient replication assay developed initially by Xie *et al.* (58), for wheat dwarf geminivirus work, and adapted by us for work with maize streak-related geminiviruses (ref. 56; J. Willment and E.P. Rybicki, unpublished). This uses BMS suspension-cultured (or less often, callus-cultured) cells that are biolistically infected with dimerized or partly-dimerized MSV genomes, or replicatable constructs based on MSV, followed by nucleic acid hybridization for detection of replicating forms. There is no obvious reason why infectious cDNA clones of RNA genomes could not be used, or even ssRNA genomes or whole virus particles. The one drawback to the technique is that it is very high-tech in its requirements, and only a small minority of cells are 'hit', meaning virus products are at far lower concentration than would be the case in a protoplast system, where a much greater proportion of cells can be transformed or infected. Moreover, it is unwise to allow cells to multiply, as the infected ones may well be diluted out if they are less viable. The advantage is that suspension (or callus) cells are much more metabolically similar to cells in the plant than are protoplasts.

Protocol 9. Transient replication assay in BMS cells

Equipment and reagents[a]

- Whatman #4 filter paper disks (7 cm diameter)
- Vacuum Buchner funnel
- BMS solid medium (2SMS2D; *Protocol 4*) containing 0.2 M mannitol and 10 mg ml^{-1} $AgNO_3$
- Tris-EDTA (10 mM Tris-Cl, 10 mM EDTA, pH 7.8)
- 5 M NaCl
- 5 M NaAc (sodium acetate)
- TE buffer (10 mM Tris-HCl; 1 mM EDTA, pH 7.5)
- 10% (w/v) sodium dodecyl sulfate (SDS)
- 96% ethanol
- Yeast tRNA stock, 10 mg ml^{-1} in water
- Proteinase K stock, 20 mg ml^{-1} in water
- Colloidal gold suspension
- Liquid nitrogen
- Microcentrifuge tube pestles
- Celite
- DuPont (BioRad) PDS-1000/He Biolistic Particle Delivery System or equivalent device
- Friable, rapidly growing BMS suspension cultured cells (*Protocol 6*)

A. *Preparation of cells for bombardment*

Caution: work aseptically!

1. Place 1 ml of suspended BMS cells on the centre of a sterile filter disk in a Buchner funnel, with a slight vacuum applied. Spread the cells evenly so as to get a single layer in an approximately 3 cm diameter circle.

2. Transfer the filter(s) to BMS solid medium for 4 h. This is a high osmoticum medium which dehydrates the cells to reduce damage due to the bombardment.

B. *Particle preparation and biolistic bombardment*

1. Precipitate plasmids onto gold particles by method of Dunder *et al.* (51). Up to 6 μg of plasmid can be used per precipitation.
2. Bombard cells under appropriate conditions: 650 p.s.i. rupture disks, a gap distance of 6 mm, macrocarrier flight distance of 5 mm, and particle flight distance of 6 cm are used by us with the BioRad gun. Each plate of tissue should be bombarded twice. 200 ng of plasmid DNA is used per shot. 200 ng each of anthocyanin regulatory genes p35SB-Peru and p35SC1 can be used per shot, to evaluate the efficiency of the bombardment experiment. Cells containing the DNA turn red due to anthocyanin production.
3. After 16 h, transfer filters to fresh BMS solid medium without mannitol. Mannitol keeps the cells partially dehydrated, which increases their survival potential after bombardment damage, as they are less liable to lysis.

C. *Isolation of DNA from callus*[b]

1. Approximately 3 days after bombardment (determine best time empirically: 3–4 days is best for MSV replicons), place disks on Whatman 3MM or other absorbent paper to absorb excess liquid, then scrape the BMS cells off the filter paper disks, and place them in a 1.5 ml microcentrifuge tube with a small amount of Celite. Freeze the tube in liquid nitrogen, and grind cells to a fine powder with a microcentrifuge tube pestle.
2. Resuspend the extract in 0.5 ml Tris-EDTA, mix gently, add 50 μl 10% SDS, and mix gently (**do not vortex**). Keep at room temp for 10 min (59).
3. Add 140 μl of 5M NaCl, and mix thoroughly but **gently** (to avoid shearing chromosomal DNA). Seal tube, keep at 4°C for at least 8 h (or overnight).
4. Centrifuge at 4°C in a minicentrifuge (e.g. 15000 r.p.m.) for 30 min. Low molecular weight DNA is in the supernatant. Transfer this to a new tube, and add 30 μg of yeast tRNA per ml, and 100 μg of proteinase K; incubate at 50°C for 1–2 h.
5. Extract twice with 1 volume of phenol-chloroform, and once with 1 volume of chloroform; separate layers by centrifugation, and keep the supernatant at all times. Precipitate DNA with 2 volumes of ethanol, at −70°C for 30 min, centrifuge at 4°C, collect pellet, and resuspend in

Protocol 9. *Continued*

400 μl of TE plus 25.5 μl of 5M NaAc. Re-precipitate with 2 volumes of ethanol.

6. Resuspend DNA pellet resulting from centrifugation in 40 μl of TE. This contains low molecular-weight DNA plus some RNA, which may be removed by treatment with RNAse A.
7. The DNA yield from each isolation is estimated by measuring the absorbance at 260 nm.
8. Electrophorese fractional or total DNA yield in agarose gel, and Southern blot to adsorptive membrane.
9. Detect replicative form nucleic acids by hybridization with a labelled genome-specific probe.

[a] All reagents should be sterile.
[b] We have also isolated virions from callus cells, and enzymatically assayed activity (such as GUS by chemiluminescence or chromogenic staining) of expressed reporter genes.

4. Assay of plant viruses

This is a topic which has been well covered over many years. The classic biological assay of plant viruses has been partially covered in *Protocol 1*. Other classic techniques are comprehensively described in ref. 60, and newer ones in ref. 61. Thus we will not cover any technique in any depth here, except for some discussion on polymerase chain reaction (PCR) methods. It suffices to say that plant virus assay techniques, like those for their animal counterparts, range from the biological through physical to the molecular biological. These include:

- local lesion and endpoint (LD_{50}, ID_{50}) symptomatic assays (*Protocol 1*)
- vector studies (transmission by insects, fungi, nematodes)
- ultraviolet spectroscopy (assay for purity)
- analytical ultracentrifugation (for particle morphology/heterogeneity)
- electron microscopy of negatively stained particles (leaf dip or immuno-capture and/or antibody decoration)
- electron microscopy of plant tissue sections, with or without immunolocalization by ferritin- or gold-labelled antibodies
- immunoprecipitin detection/assay
- dot-blot or tissue-print enzyme immunoassays on membranes
- microplate enzyme immunoassays (capture/sandwich, indirect techniques)
- 'western' or immunoelectroblotting detection
- dot-blot genomic nucleic acid hybridization with radio- or ligand-labelled probes

- Southern or northern blotting (after electrophoresis)
- *in situ* nucleic acid hybridization
- polymerase chain reaction (PCR) amplification of genomic copy DNA; this can be quantitative, for accurate estimation of amount of virus genome present

Obviously whole plant work is still very important, since this is how a virus is characterized biologically. However, techniques such as dot-blotting and double antibody sandwich enzyme-linked immunosorbent assay (DAS-ELISA) have been routinely used around the world for some years now for detection and assay of plant viruses. These uses are both commercial, and for purposes of plant protection and phytosanitary protocols.

4.1 PCR amplification

In recent years, the use of PCR has become very popular. A recent query to the NCBI PubMed literature database (http://www.ncbi.nlm.nih.gov/PubMed/) turned up over 70 citations to the query statement 'PCR plant virus detection', in only the last eight years. The techniques in use range from simple use of sequence-specific primers for DNA amplification from Geminiviridae or Caulimoviruses, to single-tube immunocapture cDNA amplification protocols for Potyviridae amplification. Protocols vary too much to give any generally useful method here; however, a general description of how to design primers, and structure DNA or cDNA amplification protocols, may be found at http://www.uct.ac.za/microbiology/pcr.htm. Certain important considerations are given below:

- Use plugged pipette tips: prevents aerosol contamination of pipettes
- DMSO often allows better denaturation of longer target sequences (>1 kb) and more product
- Do not use the same pipette for dispensing nucleic acids that you use for dispensing reagents
- Remember sample volume should not exceed one-tenth of the reaction volume, and sample DNA/NTP/primer concentrations should not be too high, since otherwise all available Mg^{2+} is chelated out of solution, and enzyme reactivity is adversely affected. Any increase in dNTPs over 200 μM means [Mg^{2+}] should be re-optimized.
- Avoid using EDTA-containing buffers, as EDTA chelates Mg^{2+}
- Low concentrations of primer, target, Taq polymerase, and nucleotide are to be favoured, as these generally ensure cleaner product and lower background, perhaps at the cost of detection sensitivity.
- Initial denaturation at start: 92–97 °C for 3–5 min. If you denature at 97 °C, denature sample only; add rest of mix after reaction cools to annealing temperature (prevents premature denaturation of enzyme).

- Initial annealing temperature: as high as feasible for 3 min (e.g. 50–75 °C). Stringent initial conditions mean less non-specific product, especially when amplifying from eukaryotic genomic DNA.
- Initial elongation temperature: 72 °C for 3–5 min. This allows complete elongation of product on rare templates.

Temperature cycling:

- 92–94 °C for 30–60 s (denature)
- 37–72 °C for 30–60 s (anneal)
- 72 °C for 30–60 s (elongate) (60 s per kb target sequence length)
- 25–35 cycles only (otherwise enzyme decay causes artefacts)
- 72 °C for 5 min at end to allow complete elongation of all product DNA

Note:

- 'Quickie' PCR is quite feasible: e.g. [94 °C 30 s/45 °C 30 s/72 °C 30 s] × 30, for short products (200–500 bp).
- Don't run too many cycles: if you don't see a band with 30 cycles you probably won't after 40; instead, take an aliquot from the reaction mix and re-PCR with fresh reagents.

and always remember:

- Work cleanly
- Don't work with PCR products in PCR preparation areas
- *Always* include water and very dilute positive controls in every experiment
- Wear gloves

Acknowledgements

I thank Donald Solomons for his wisdom concerning plants, and my students for sharing their protocols.

References

1. Mayo, M. A., and Pringle, C.R. (1998). *J. Gen. Virol.,* **79**, 649.
2. Pringle, C. R. (1998). *Arch. Virol.,* **143**, 203.
3. Mullineaux, P. (1992). In *Molecular plant pathology: a practical approach.* (ed. S. J. Gurr, M. J. McPherson, and D. J. Bowles), p. 11. Oxford University Press, Oxford.
4. Kroner, P., and Ahlquist, P. (1992). In *Molecular plant pathology: a practical approach.* (ed. S. J. Gurr, M. J. McPherson, and D. J. Bowles), p. 23. Oxford University Press, Oxford.
5. Francki, R. I. B. (1972). In *Principles and techniques in plant virology.* (ed. C. I. Kado, and H. O. Agrawal), p. 295. Van Nostrand Reinhold, NY.

6. Hull, R. (1992). In *Molecular plant pathology: a practical approach.* (ed. S. J. Gurr, M. J. McPherson, and D. J. Bowles), p. 1. Oxford University Press, Oxford.
7. Holmes, F. O. (1929). *Botanical Gazette,* **87**, 39.
8. Dulbecco, R. (1952). *Proc Natl Acad Sci USA,* **38**, 747.
9. Ehrig, F., Golovcenko, O., and Schmidt, H. E. (1990). *Arch. Phytopathol. Pflanzenschutz.,* **26** , 105.
10. Jakab, G., Droz, E., Brigneti, G., Baulcombe, D., and Malnoe, P. (1997). *J. Gen. Virol.,* **78**, 3141.
11. Louie, R. (1995). *Phytopathology,* **85**, 139.
12. *Plant virology protocols: from virus isolation to transgenic resistance* (ed. G. Foster, and S. Taylor) (1998). Humana Press, Totowa, NJ, p. 571.
13. Rybicki, E. P., and von Wechmar, M. B. (1981). *Virology,* **109**, 391.
14. Murant, A. F. and Harrison, B. D. (1970). *Descriptions of plant viruses.* Commonwealth Mycological Institute and the Association of Applied Biologists, Kew, England.
15. van Regenmortel, M. H. V. (1972). *Principles and techniques in plant virology.* (ed. C. I. Kado, and H. O. Agrawal), p. 390. Van Nostrand Reinhold, NY.
16. Williamson, C., Rybicki, E. P., Kasdorf, G. G. F., and von Wechmar, M. B. (1988). *J. Gen. Virol.,* **69**, 787.
17. Dodds, J. A., Morris, T. J., and Jordan, R. L. (1984). *Annu. Rev. Phytopathol.,* **22**, 151.
18. Valverde, R. A. (1990). *Plant Disease,* **74**, 255.
19. Covey, S. N., Noad, R. J., Al-Kaff, N. S., and Turner, D. S. (1998). In *Plant virology protocols: from virus isolation to transgenic resistance* (ed. G. Foster, and S. Taylor), p. 53. Humana Press, Totowa, NJ.
20. Palmer, K. E., Schnippenkoetter, W. H., and Rybicki, E. P. (1998). In *Plant virology protocols: from virus isolation to transgenic resistance* (ed. G. Foster, and S. Taylor), p. 41. Humana Press, Totowa, NJ.
21. Gilbertson, R. L., Faria, J. C., Hanson, S. F., Morales, F. J., Ahlquist, P., Maxwell, D. P., and Russell, D. R. (1991). *Phytopathology,* **81**, 980.
22. Wyatt, S. D., Druffel, K., and Berger, P. H. (1993). *J. Virol. Methods,* **44**, 211.
23. Berger, P. H., and Shiel, P. J. (1998). In *Plant virology protocols: from virus isolation to transgenic resistance* (ed. G. Foster, and S. Taylor), p. 151. Humana Press, Totowa, NJ.
24. Dodds, J. A. (1986). In *Development and applications in virus testing.* (ed. R. A. C. Jones, and L. Torrance), p. 71. The Lavenham Press, London.
25. Gehringer, M. (1996). *The construction of a virus expression vector for the high level production of proteins in plants.* M.Sc. Thesis, University of Cape Town.
26. Hall, S. (1986). *The use of double-stranded RNA in the identification and characterization of cucumber mosaic virus.* B.Sc. Hons. Thesis, University of Cape Town.
27. Hughes, F. L. (1991). *Molecular investigations of subgroup I geminiviruses.* Ph.D. Thesis, University of Cape Town.
28. Gal-On, A., Meiri, E., Elman, C., Gray, D. J., and Gaba, V. (1997). *J. Virol. Methods,* **64**, 103.
29. Franche, C., Bogusz, D., Schopke, C., Fauquet, C., and Beachy, R. N. (1991). *Plant Mol. Biol.,* **17**, 493.
30. Sambrook, J., Fritsch, E. F., and Maniatis, T. (1989). *Molecular cloning: a laboratory manual* (2nd edn), p. B.4. Cold Spring Harbor Laboratory, NY.

31. Briddon, R. W., Prescott, A. G., Lunness, P., Chamberlin, L. C., and Markham, P. G. (1993). *J. Virol. Methods,* **43**, 7.
32. Brugidou, C., Holt, C., Ngon, A., Yassi, M., Zhang, S., Beachy, R., and Fauquet, C. (1995). *Virology,* **206**, 108.
33. Fakhfakh, H., Vilaine, F., Makni, M., and Robaglia, C. (1996). *J. Gen. Virol.,* **77**, 519.
34. Lamprecht, S., and Jelkmann, W. (1997). *J. Gen. Virol.,* **78**, 2347.
35. Bajaj, Y. P. S. (1994). In *Maize* (ed. Y. P. S. Bajaj), p. 3. Springer-Verlag, Wien.
36. Potrykus, I., and Spangenberg, G. (1995). *Gene transfer to plants.* Springer-Verlag, Berlin.
37. Hansen, A. J. (1974). *Virology,* **57**, 387.
38. Moskovets, S. N., Didenko, L. F. and Zhuk, I. P. (1971). *Mikrobiol. Zh.,* **33**, 578.
39. Moskovets, S. N., Zhuk, I. P., Didenko, L. F., and Gorbarenko, N. I. (1971). *Vopr. Virusol.,* **16**, 430.
40. Bornman, C. H. (1993), p. 246. In *Plant breeding: principles and prospects* (ed. M. D. Hayward, N. O. Bosemark, and I. Romagosa), p. 246. Chapman and Hall, London.
41. George, E. F., and Sherrington, P. D. (1983). *Plant propagation by tissue culture. Handbook and directory of commercial laboratories.* Exegetics, Edinton.
42. Kyte, L. (1998). *Plants from test tubes: an introduction to micropropagation.* Timber Press, Oregon.
43. George, E. F., Puttock, D. J. M., and George, H. J. (1987). *Plant Culture Media,* Vol 1. *Formulations and uses.* Exegenetics, Edington.
44. Freeling, M., and Walbot, V. (1994). In *The maize handbook.* (ed. M. Freeling, and V. Walbot), p. 663. Springer-Verlag, New York.
45. Russell, D. A., and Fromm, M. E. (1995). In *Gene transfer to plants* (ed. I. Patpykus and G. Spangenbern), p. 118. Springer-Verlag, Berlin.
46. Kirihara, J. A. (1994). In *The maize handbook.* (ed. M. Freeling, and V. Walbot), p. 690. Springer-Verlag, New York.
47. Armstrong, C. L. (1994). In *The maize handbook.* (ed. M. Freeling, and V. Walbot), p. 663. Springer-Verlag, New York.
48. Murashige, T., and Skoog, F. (1962). *Physiol. Plant.,* **15**, 473.
49. Armstrong, C. L., Green, C. E., and Phillips, R. L. (1991). *Maize Genetics Corporation Newsletter,* **65**, 92.
50. Sellmer, J. C., Ritchie, S. W., Kim, I. S., and Hodges, T. K. (1994). In *The maize handbook.* (ed. M. Freeling, and V. Walbot), p. 671. Springer-Verlag, New York.
51. Dunder, E., Dawson, J., Suttie, J., and Pace, G. (1995). In *Gene transfer to plants* (ed. I. Potrykus, and G. Spangenberg), p. 126. Springer-Verlag, Berlin.
52. Chu, C. C., Wang, C. C., Sun, C. S., Hus, C., Yin, H. C., and Chu, C. Y. (1975). *Scientia Sinica,* **18**, 659.
53. Klein, T. M. (1994). In *Maize* (ed. Y. P. S. Bajaj), p. 241. Springer-Verlag, Wien.
54. Shillito, R. D., Carswell, G. K., and Kramer, C. (1994). In *The maize handbook.* (ed. M. Freeling, and V. Walbot), p. 695. Springer-Verlag, New York.
55. Di Maio, J. J., and Shillito, R. D. (1989). *J. Tissue Culture Meth.,* **12**, 163.
56. Palmer, K. E. (1997). *Investigations into the use of maize streak virus as a gene vector.* Ph.D. Thesis, University of Cape Town.

57. Muller, D. G., Sengco, M., Wolf, S., Braughtigan, M., Schmid, C., Knapp, M., and Kruppers, R. (1996). *J. Gen. Virol.,* **77**, 2329.
58. Xie, Q., Suarez-Lopez, P., and Gutierrez, C. (1995). *EMBO J.,* **14**, 4073.
59. Anat, S., and Subramanian, K. N. (1992). *Methods in Enzymology* Vol. 216, p. 20. Academic Press, London.
60. Kado, C. I., and Agrawal, H. O. (1972). *Principles and techniques in plant virology.* Van Nostrand Reinhold, NY.
61. Jones, R. A. C., and Torrance, L. (1986). *Developments and applications in virus testing.* Vol. 1. Association of Applied Biologists, Wellesbourne, UK.

A1

List of suppliers

Adcoa Corporation, 1269 Eagle Vista Drive, Los Angeles, CA 90041, USA. (http://www.thomasregister.com/olc/adcoa/)

Advanced Biotechnologies Ltd, Unit B1-B2, Longmead Business Centre, Blenheim Road, Epsom, Surrey KT19 9QQ, UK. (http://www.adbio.co.uk/)

Amersham Pharmacia Biotech, Amersham Place, Little Chalfont, Buckinghamshire HP7 9NA, UK. (http://www.apbiotech.com/)

American Type Culture Collection, P.O. Box 1549, Manassas, Virginia 20108, USA. (http://www.atcc.org/)

Amrad Biotech, 34 Wadhurst Drive, Boronia, Victoria 3155, Australia. (http://www.amrad.com.au/)

Anachem Ltd, 20 Charles Street, Luton, Bedfordshire LU2 0EB, UK. (http://www.anachem.co.uk/)

Authentikit System, Innovative Chemistry, Marshfield, MA, USA.

BDH Laboratory Supplies, Poole, Dorset BH15 1TD, UK. (http://www.bdh.com/)

Beckman Instruments (UK) Ltd, Oakley Court, Kingsmead Business Park, London Road, High Wycombe, Buckinghamshire HP11 1JU, UK.

Beckman Coulter Inc., 4300 N. Harbor Boulevard, P.O. Box 3100, Fullerton, CA 92834-3100, USA. (http://www.beckman.com/)

Becton Dickinson, 1 Becton Drive, Franklin Lakes, New Jersey 07417-1883, USA. (http://www.bd.com/)

Bibby Sterilin Ltd, Tilling Drive, Stone, Staffordshire ST15 0SA, UK.

BioMérieux SA, 69280 Marcy l'étoile, France. (http://www.biomerieux.fr/)

Bio-Rad Laboratories Ltd, Bio-Rad House, Maylands Avenue, Hemel Hempstead, Hertfordshire HP2 7TD, UK.

Bio-Rad Laboratories, Life Science Research, 2000 Alfred Nobel Drive, Hercules, California 94547, USA. (http://www.biorad.com/)

Boehringer Mannheim, see Roche

Carl Zeiss Ltd, 17-20 Woodfield Road, P.O. Box 78, Welwyn Garden City, Hertfordshire AL7 1LU, UK.

Carl Zeiss Inc., One Zeiss Drive, Thornwood, NY 10594, USA. (http://www.zeiss.com/)

Celite Corporation, 137 West Central Ave, Lompoc, CA 93436, USA.

Chiron Corporation Headquarters, 4560 Horton St, Emeryville, CA 94608-2916, USA. (http://www.chiron.com/)

Corning Incorporated, Science Products Division, 45 Nagog Park, Acton, MA 01720, USA. (http://www.corningcostar.com/)

Costar, see Corning

Creg Veterinary Foods, 3 Bromwich Ave, Highgate, London, N6 6QH, UK. (http://www.liquivite.mcmail.com/)

DAKO Ltd, Denmark House, Angel Drove, Ely, Cambridgeshire CB7 4ET, UK.

Dako Corporation, 6392 Via Real, Carpinteria, CA 93013 USA. (http://www.dako.com/)

Denley Labsystems, Affinity Sensors, Saxon Way, Bar Hill, Cambridge CB3 8SL, UK.

Denley Labsystems Inc., 8 East Forge Parkway, Franklin, MA 02038, USA. (http://www.denley.com/)

Deutsche Sammlung von Mikroorganismen und Zellkulturen GmbH, Mascheroder Weg 1B, D-3300 Braunschweig, Germany. (http://www.dsmz.de/)

Dynalab Corporation, PO Box 112, Rochester, NY 14601-0112, USA. (http://www.bibby-sterilin.com/)

Dynex Technologies, Daux Road, Billingshurst, West Sussex RH14 9SJ, UK.

Dynex Technologies, 14340 Sullyfield Circle, Chantilly, VA 20151-1683, USA. (http://www.dynextechnologies.com/)

Edwards High Vacuum International, West Sussex, UK.

Electron Microscopy Sciences, P.O. Box 251, 321 Morris Road, Fort Washington, PA 19034, USA. (http://www.emsdiasum.com/ems/)

Eurogenetics UK, Unit 5, Kingsway Business Park, Oldfield Road, Hampton TW12 2HD, UK. (http://www.eurogenetics.be/)

European Collection of Cell Cultures, Centre for Applied Microbiology & Research, Salisbury, Wiltshire SP4 0JG, UK. (http://www.camr.org.uk/ecacc.htm/)

Flowgen, Lynn Lane, Shenstone, Staffs WS14 0EE, UK. (http://www.flowgen.co.uk/)

Fluka, see Sigma-Aldrich.

Gallenkamp, see Sanyo-Gallenkamp.

Gilson Company Inc., P.O. Box 677, Worthington, Ohio 43085-0677, USA. (http://www.globalgilson.com/)

Grant Instruments (Cambridge) Limited, Barrington, Cambridge CB2 5QZ, UK. (http://www.grant.co.uk/)

Harlan Sera-Lab Ltd, Dodgeford Lane, Belton, Loughborough, Leicestershire LE12 9TE, UK. (http://www.harlan.com/)

Heraeus Instruments, 9 Wates Way, Ongar Road, Brentwood, Essex CM15 9TB, UK.

Heraeus Instruments, 111-A Corporate Blvd, South Plainfield, NJ 07080, USA. (http://www.heraeus-instruments.com/)

ICN Biomedicals, 1 Elmwood, Chineham Business Park, Crockford Lane, Basingstoke, Hampshire RG24 8WG, UK.

ICN Pharmaceuticals Inc., 3300 Hyland Avenue, Costa Mesa, CA 92626, USA. (http://www.icnbiomed.com/)

Labtrack, P.O. Box 19, Uckfield, East Sussex GN22 3TS, UK.

Leica Microsystems (UK) Ltd, Davy Avenue, Knowlhill, Milton Keynes MK5 8LB, UK.

Leica Microsystems Inc., 111 Deer Lake Road, Deerfield, IL 60015, USA. (http://www.leica.com/)

Leitz, see Leica Microsystems.

Life Technologies (Gibco BRL), 9800 Medical Center Drive, Rockville, MD 20850, USA. (http://www.lifetech.com/)

Macherey-Nagel GmbH & Co, Postfach 10 13 52, D-52313 Düren, Germany.

Macherey-Nagel Inc., 6 South Third St., Suite 402, Easton, PA 18042, USA. (http://www.macherey-nagel.com/)

Merck & Co. Inc., Whitehouse Station, NJ, USA. (http://www.merck.com/)

Millipore (UK) Ltd, The Boulevard, Blackmoor Lane, Watford WD1 8YW, UK.

Millipore Corporation, 80 Ashby Road, Bedford, Massachusetts 01730-2271, USA. (http://millispider.millipore.com/)

Nalge Nunc (Europe) Limited, Foxwood Court, Rotherwas Industrial Estate, Hereford HR2 6JQ, UK.

Nalge Nunc International, 75 Panorama Creek Drive, Rochester, NY 14625, USA. (http://nunc.nalgenunc.com/)

Nikon UK Ltd, 380 Richmond Road, Kingston upon Thames, Surrey KT2 5PR, UK.

Nikon Inc., 1300 Walt Whitman Road, Melville, NY 11747-3064, USA. (http://www.nikon.com/)

Novartis, http://www.novartis.com/

Noveargis Pharmaceuticals UK Ltd (makers of Amantadine), Wimelehueld Road, Horsham, West Sussex, RH12 4AB, UK.

Organon Teknika, Boseind 15, 5281 RM Boxtel, The Netherlands. (http://www.organonteknika.com/)

Oxoid Limited, Wade Road, Basingstoke, Hampshire RG24 8PW, UK. (http://www.oxoid.co.uk/)

Philip Harris Scientific, 618 Western Avenue, Park Royal, London W3 0TE, UK. (http://www.philipharris.co.uk)

Polysciences Inc., 400 Valley Road, Warrington, PA 18976, USA. (http://www.polysciences.com/)

Promega UK, Delta House, Enterprise Road, Chilworth Research Centre, Southampton SO1 7NS, UK.

Promega Corporation, 2800 Woods Hollow Road, Madison WI 53711, USA. (http://www.promega.com/)

Qiagen Ltd, Boundary Court, Gatwick Road, Crawley, West Sussex RH10 2AX, UK.

Qiagen Inc., 28159 Avenue Stanford, Valencia, CA 91355, USA. (http://www.qiagen.com/)

Riken Cell Bank, 3-1-1 Koyadai, Tsukuba Science City, 305 Iboraki, Japan. (http://www.rtc.riken.go.jp/)

Roche: F. Hoffmann-La Roche Ltd, CH-4070 Basel, Switzerland. (http://www.roche.com/)

Sandoz, see Novartis.

Sanyo-Gallenkamp, Riverside Way, Uxbridge, Middlesex UB8 2YF, UK. (http://www.sanyo.co.uk/gallenkamp/)

Sartorius Ltd, Longmead Business Centre, Blenheim Road, Epsom, Surrey KT19 9QN, UK.

Sartorius North America Inc., 131 Heartland Blvd, Edgewood, New York 11717, USA. (http://www.sartorius.com/)

Sera-Lab, see Harlan Sera-Lab.

Seward Ltd, 98 Great North Road, London N2 0GN, UK. (http://www.seward.co.uk/)

Sigma-Aldrich Company Ltd, Fancy Road, Poole, Dorset BH12 4QH, UK

Sigma, 3050 Spruce St, St. Louis, P.O. Box 14508, St. Louis MO 63178, USA. (http://www.sigma-aldrich.com/)

Worthington Biochemical Corporation, Lorne Laboratories Ltd, 7 Tavistock Estate, Ruscombe Business Park, Ruscombe Lane, Twyford, Reading, Berkshire RG10 9NJ, UK. (http://www.worthington-biochem.com/)

Zeiss, see Carl Zeiss.

Index